D1765671

0784401225

FOREWORD

The activities of the Committee on Education of the ASCE, Technical Council on Forensic Engineering are aimed at encouraging the inclusion of forensic engineering topics and failure case studies in civil engineering education. The development of resource materials for undergraduate, graduate and continuing professional education is a principal ongoing project of this committee. The object of this Special Publication is to assemble in one place a convenient source of typical case studies, including an outline of each, a summary of the lessons learned and a list of background references for further study.

Failures typically occur during the construction of a facility as well as during the service life, often when subjected to an unusual or unanticipated load situation. Some of the failures may be classified as "milestone" in that one single event resulted in a decisive change to design or construction practices. Others may be categorized as "last straw" events in that one is selected to represent an endemic problem which was only addressed effectively when a continuing problem reached such significance that a serious effort was made to tackle it. It is recognized that "failure" may involve loss of serviceability as distinct from collapse and that failure is an extreme form of damage which itself constitutes a material, nontrivial change in the safety, serviceability, appearance or repairability of the constructed facility.

It is anticipated that this document will provide a starting point for the study of the performance and failure of a variety of civil engineering structural, foundation and geoenvironmental systems. It is hoped that its availability will encourage more emphasis being placed on such studies, with consequent improved awareness of the fact that much sound engineering benefits not only from a knowledge of past successes but of past failures as well.

The nature of civil engineering construction differs from that of almost all other manufacturing processes in that, most often, only one product is produced on each occasion. The opportunity to build a series of mock-ups and to improve progressively on sequential attempts at one project is rarely, if ever, available to Civil Engineers. Additionally, the regular yet often unpredictable nature of the exposure of constructed facilities to natural hazards provides particular difficulties in selecting acceptable design criteria, and in assessing the quality of the performance of structures under extreme conditions.

These factors present a particularly imposing challenge to those involved in the design, construction and operation of constructed facilities and result in continuing benefit being derived from the ongoing Learning from Failure process. Although failures may be infrequent, the impact is often devastating and may give rise to vociferous, even if misinformed, criticism of those involved in all aspects of the project. A comprehensive awareness of the experiences of the past will assist in developing in the engineers of the present and future a critical awareness of the need to be ever vigilant in the search for safe and successful procedures. It is the hope of those who have assembled this summary of case studies that it will serve to prompt an awareness of the value of studying past failures and to encourage further interest in the field.

ACKNOWLEDGMENTS

A number of individuals helped in the preparation of the Special Publication. The secretarial assistance of Ms. Vicki Clopton is gratefully acknowledged. Appreciation is also extended to Mr. Martin Stark Rendon, whose interpretation of the failure of arguably the most infamous suspension bridge ever, the Tacoma Narrows Suspension Bridge ("Galloping Gertie"), is depicted on the cover of this publication.

Contributors

The short descriptions contained in this publication were contributed by the following individuals:

Paul A. Bosela, Cleveland State University, Cleveland, Ohio

Kenneth L. Carper, Washington State University, Pullman, Washington

Timothy J. Dickson, Construction Technology Laboratories Inc., Skokie, Illinois

J. David Frost (Co-Editor), The Georgia Institute of Technology, Atlanta, Georgia

Narbey Khachaturian, The University of Illinois, Champaign-Urbana, Illinois

Oswald Rendon-Herrero, Mississippi State University, Starkville, Mississippi

Robin Shepherd (Co-Editor), Forensic Expert Advisers Inc., Santa Ana, California

Robin Shepherd	J. David Frost
Co-Editor	Co-Editor

CONTENTS

Building Failures

FOUNDATION FAILURES

TOWER OF PISA - (1173 & Ongoing)

The Tower of Pisa in Italy was constructed in three phases. Four floors were built over a 5 year period from 1173 to 1178. Following an almost 100 year hiatus, three additional floors were constructed between 1272 and 1278. The third construction phase occurred more than 80 years later between 1360 and 1370 when the bell tower was added.

Evidence indicates that the phased construction was mandated by the performance of the structure as it was being built. Further, records indicate that during construction, the tower appeared to move sufficiently so that the builders used obliquely cut stones in an effort to maintain the floor of each successive story approximately horizontal. It is interesting to note that the obliquely cut stones were used by the Italian Commission, that was entrusted with gathering and collating relevant data for an international competition organized to identify a method to stabilize the tower in 1972, to reconstruct the pattern of movements throughout the first two phases of construction of the tower. These calculations showed that at the end of the first construction phase the tower had begun to lean towards the Northwest. During the second and third phases, the angle of inclination increased and the principal direction of tilt shifted first to the Northeast and then to the South.

Currently the tower which is about 60 m (200 ft) tall, 20 m (66 ft) in diameter and weighs approximately 145 MN (14,500 tons) is inclined at almost 5.5 degrees to the South or about 6 m (20 ft) out of plumb. The first recorded modern direct measurement of the angle of inclination was in 1911. Measurements made since then have shown that when no external forces/conditions are acting, the rate of inclination increases approximately at a constant rate. For example, from mid 1975 to mid 1985 when external disturbances were at a minimum, the average rate of inclination remained approximately constant at about 5.8 seconds of arc per year. Conversely drawdown of groundwater during the period late 1970 to late 1974 caused a dramatic increase in inclination rate to 11.9 seconds of arc per year. Similar increases have accompanied other invasive construction activities over the past 80 years.

In June, 1993, a scheme to reduce the tilt of the tower using specially fabricated lead weights placed on top of the North side of a tensioned concrete ring built around the base of the tower was initiated. The tilt was reduced by approximately 12 seconds of arc over a 6 week period when about 1.3 MN (130 tons) of lead had been placed. At the time of writing, this remedial program continues as additional weights, up to an initial design amount of 6.9 MN (690 tons), are added.

Lessons Learned:

The failure of the Tower of Pisa is without doubt unique for a number of reasons ranging from the fact that it is a failure that has been occurring essentially on a continuous basis for more than 800 years. Despite the extensive investigations and analyses conducted over the past 60 years, there is still no consensus on the cause of failure. What is significant however is that finally, after 8 centuries, the condition of the tower has been improved in a controlled fashion.

References:

Mitchell, J., K., Vivatrat, V., and Lambe, T.W., (1977), "Foundation Performance of the Tower of Pisa", ASCE, Journal of Geotechnical Engineering Division, Vol. 103, No. GT 3, pp. 227-247.

Leonards, G.A., (1979), Discussion of "Foundation Performance of the Tower of Pisa", ASCE, Journal of Geotechnical Engineering Division, Vol. 105, No. GT1, pp. 95-105.

Jamiolkowski, M., (1994), "Leaning Tower of Pisa - Description of the Behavior", Settlement '94 Banquet Lecture, College Station, Texas, 55 pp.

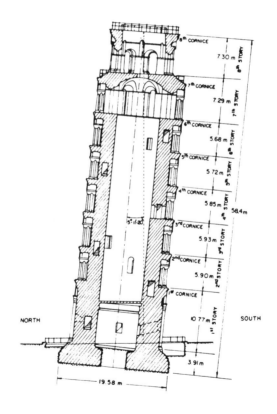

Section through the Tower of Pisa
(Mitchell et al., 1977; Reprinted by permission of ASCE)

TRANSCONA GRAIN ELEVATOR - (1913)

Construction for a million-bushel grain elevator began in 1911 at North Transcona, near Winnipeg, Manitoba, Canada. The elevator structure consisted of a reinforced concrete work-house and an adjoining bin-house. In plan, the work-house measured 21 by 29 m (70 by 90 ft). The structure was 55 m (180 ft) high and was founded on a raft foundation 3.6 m (12 ft) below grade. The bin-house consisted of five rows of thirteen bins each 4.3 m (14 ft) in diameter and 28 m (92 ft) high which rested on a concrete framework supported by a reinforced-concrete raft. The bin-house raft measured 23 m (77 ft) by 59 m (195 ft) and was also founded at a depth of 3.7 m (12 ft) below grade.

Upon completion in September 1913, filling of the elevator commenced and the grain was distributed as evenly as possible amongst the bins. On October 18, 1913, after 31,500 cu m (875,000 bushels) of wheat had been placed in the elevator, settling was noted which increased uniformly to about 0.3 m (1 ft) per hour. Tilting of the elevator then began to occur which ceased 24 hours later when the inclination reached 26 degrees 53 minutes from vertical.

The subsoil below the elevator foundation consisted of a uniform deposit of clay which was the result of sedimentation in waters of glacial Lake Agassy. The clay was a varved slickensided highly plastic material varying from about 9 to 15 m (30 to 50 ft) in depth. It overlay a glacial till on a limestone bedrock. From the ground surface to a depth of about 9 m (30 ft), the clay had a stiff consistency which gradually decreased just the glacial till. A water table was encountered at a depth of about 9 m (30 ft).

In 1951, a comprehensive geotechnical investigation led to the conclusion that the elevator foundation had failed due to a bearing failure in the underlying clay. Unfortunately for the design engineers, the state-of-the-art in geotechnical engineering in 1911 had not reached the point that the ultimate bearing capacity could be computed. No borings were known to have been made for the design of the elevator.

Lessons Learned:

The development of soil mechanics after the Transcona failure eventually provided a basis for computing the ultimate bearing capacity of soils. It was subsequently realized, therefore, that the Transcona failure served as a "full-scale" check of the validity of such computations. In hindsight, had the Transcona engineers had access to soil mechanics theory, the failure could have been averted.

References:

Allaire, A., (1916), "The Failure and Righting of a Million-Bushel Grain Elevator," Transactions, ASCE, Vol. LXXX, pp. 800-803.

Peck, R.B. and Bryant, F.G., (1953), "The Bearing-Capacity Failure of the Transcona Elevator," Geotechnique, Vol. III, pp. 201-208.

White, L.S., (1953), "Transcona Elevator Failure: Eye-Witness Account," Geotechnique, Vol. III, p. 209.

FARGO GRAIN ELEVATOR - (1956)

Beginning in the summer of 1954, a reinforced concrete grain elevator was constructed in a level field near Fargo, North Dakota. During the autumn and the winter, a small amount of grain was placed in the elevator. Major filling of the elevator did not begin until the latter part of April 1955. On the morning of June 12, 1955, the elevator collapsed and was completely destroyed. The collapse of the structure was a classic example of a full-scale bearing capacity failure.

The grain elevator was a reinforced concrete structure consisting of twenty circular bins, twenty-six small interstitial bins at one end, with a combined bin and work house, at the other end of the structure. The structure in plan view is a long rectangle-shaped area measuring 16 by 67 m (52 by 218 ft) overall. The structures' foundation was a reinforced concrete raft 0.7 m (2.3 ft) in thickness. The bottom of the raft foundation was located 1.8 m (6 ft) below grade. Sheet piles lined the raft periphery and were thought to be driven to a depth of 5.5 m (18 ft). The elevator was founded on the fine grained sediments of Old Lake Aggassiz.

A study including a subsoil investigation and laboratory testing was conducted to determine if the failure had resulted from inadequate bearing capacity of the subsoil. The results of the study indicated that overstressing of the subsoil was the reason for the failure. Investigation of full-scale foundation failures are rare; like the Transcona grain elevator failure, this case also affords the opportunity to assess the state-of-the-art in bearing capacity analysis.

Lessons Learned:

The Fargo grain elevator collapse is a classic example of a full-scale bearing capacity failure. The failure has provided very unique and useful bearing capacity data for assessing the validity of the state-of-the-art analytical procedures.

References:

Deere, D.H., and Davisson, M.T., (1965)," Behavior of Grain Elevator Foundations Subjected to Cyclic Loading," Sixth Inter. Conf. on Soil Mechanics and Foundation Engineering, Montreal, Vol. , pp. .

Nordlund, R.L. and Deere, D.H., (1970), "Collapse of Fargo Grain Elevator," Journal of the Soil Mechanics and Foundation Engineering Division, ASCE, Vol. 96, No. SM2, pp. 585-607.

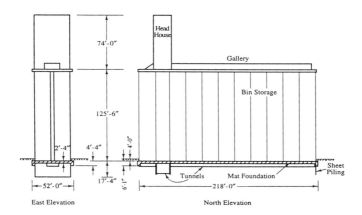

Elevation views of Fargo Grain Elevator
(Nordlund and Deere, 1970; Reprinted by permission of ASCE)

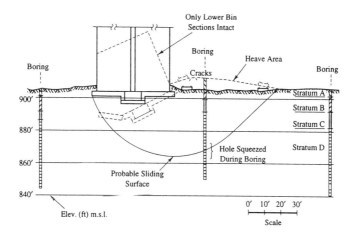

Cross-section of Collapsed Fargo Grain Elevator
(Nordlund and Deere, 1970; Reprinted by permission of ASCE)

LA PLAYA GUATEMALA EARTHQUAKE - (1976)

On February 4, 1976, an earthquake of 7.5 magnitude on the Richter scale, devastated the vacation community of La Playa, Amatitlan Lake, Guatemala. The earthquake induced liquefaction in the subsoils underlying La Playa. The earthquake's epicenter was located 170 km (106 miles) northeast of Lake Amatitlan. Within the zone of heavy damage, there was subsidence and flooding of beach areas and severe ground cracking which resulted in severe damage to houses as well as numerous sand boils. Of 32 houses at La Playa, 29 were destroyed or damaged, generally as the result of lateral ground spreading and subsidence. Most of the vacation houses were unoccupied at the time of the earthquake. A comprehensive study was conducted including field and laboratory sampling and testing.

The results of the study provided a very useful case history in which field data on soil characteristics in an earthquake-liquefied zone and non-liquefied zone could be correlated with field performance. These results supplemented the limited number of available case studies of this type which could be used for the predictions of probable behavior at other sites and permitted corroboration of analytical procedures.

Lessons Learned:

The results of the field studies from La Playa contribute to the available data base relating to earthquake-induced liquefaction and thus improve the predictive capability in earthquake-prone region.

References:

Espinosa, A.F., (1976), "The Guatemala Earthquake of February 4, 1976, A Preliminary Report," Geol. Survey Paper 1002, U.S. Gov. Printing Office, Washington, D.C.

Krinitzsky, E.L. and Bonis, S.B., (1976), "Notes on Earthquake Shaking in Soils, Guatemala Earthquake of 4 February, 1976," Informal Report, U.S. Army, Washington, D.C., 32 pp.

Hoose, S.N., Wilson, R.C., and Rosenfeld, J.H., (1978), "Liquefaction-Caused Ground Failure During the February 4, 1976, Guatemala Earthquake, Proc of the Inter. Symposium on the February 4th Earthquake and the Reconstruction Process, Vol. 2.

Seed, H.B., Arango, I., Chan, C.K., Gomez-Masso, A., and Ascoli, R., (1981), "Earthquake-Induced Liquefaction Near Lake Amatitlan, Guatemala", Journal of the Geotechnical Engineering, ASCE Vol. 107, No. GT4, pp. 501-518.

SCHOHARIE CREEK BRIDGE - (1987)

The Schoharie Creek bridge was constructed in 1953 and carries the New York State throughway across the creek. It consists of five simply supported spans with a total distance between the embankments of some 165 m (540 ft). The bridge carries two lanes of traffic in each direction approximately 24 m (80 ft) above the creek bed. On April 5, 1987, during the worst flooding experienced in years, the bridge collapsed. Four cars, a truck and ten lives were lost as a result. The sudden rerouting of the highway disrupted business on both sides of the river and focused attention on the collapse of an otherwise somewhat undistinguished structure.

Investigations confirmed scour of the pier supports as the primary cause of failure. The primary defense of the Schoharie Creek bridge against scour was the use of dry rip-rap. This consisted of large field quarry stones shaped as right rectangular prisms to inhibit rolling in flood waters. Problems had been encountered during early 1955 when vertical cracks were observed in the pier plinths. A heavily reinforced concrete element 0.9 m (3 ft) thick was cast on top of each plinth in 1957 in an attempt to control further cracking. Inspections were conducted in 1983 and 1986 but on the second occasion high water prevented detailed inspections of the bottoms of the piers.

During the April 1987 flood, the scour removed the support from the southern portion of the plinth thereby generating tensile bending stresses in the top of the plinth leading to eventual fracture through the plinth and cross footing which served to connect the two support columns. An immediate loss of support between spans 3 and 4 was then inevitable.

Lessons Learned:

The failure due to scour of this bridge reemphasizes the necessity for ensuring that bridge footings are deep enough to avoid the loss of support capacity arising from scour around the foundation. Additionally, in this instance an ill-conceived repair no doubt contributed to the ultimate failure. The presence of flood waters during the 1986 inspection inhibited a thorough inspection of the bridge and, with hindsight it appears to have been irresponsible for those involved not to have reinspected when the flood had receded and it would have been possible to have undertaken a more comprehensive examination of the footings of the columns. The continuing problem with scour causing bridge collapse prompted interest in improved technology for underwater inspection.

References:

"Collapse of the New York State Thruway Bridge over the Schoharie Creek," (1987), U.S. Senate Committee on Water Resources, Transportation and Infrastructure.

"Collapse of Thruway Bridge at Schoharie Creek," (1987), Final report, New York State Thruway Authority, November.

"Lessons from Schoharie Creek," (1988), Civil Engineering, ASCE, Vol. 58, No. 5, pp. 46-49.

Swenson, D.V. and Ingraffea, A.R., (1991), "The Collapse of Schoharie Creek Bridge - A Case Study in Concrete Fracture Mechanics", International Journal of Fracture, Vol. 51, pp. 73-92.

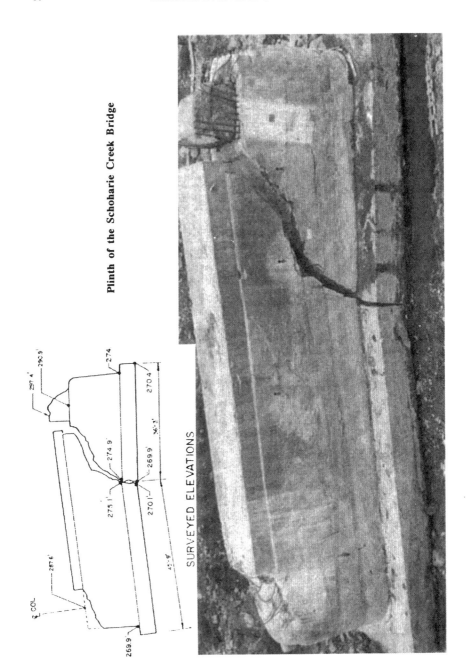

Plinth of the Schoharie Creek Bridge

EMBANKMENT, DAM AND SLOPE FAILURES

MALPASSET DAM - (1959)

The Malpasset Dam in Southern France was a double curvature arch dam with a maximum height of about 60 m (220 ft) and a crest length of about 223 m (730 ft). The thickness of the concrete varied from 1.5 m (5 ft) at the crest to 6.8 m (22 ft) at the center of the base. The dam created a reservoir with an estimated total capacity of about 51 million cu m (67 million cu yds).

The reservoir was filled very slowly over a five year period and at the time of failure, the level of the water was still 0.3 m (1 ft) below the spillway elevation. Following a period of heavy rain which resulted in the water level rising by almost 4 m (13 ft) in the three days immediately preceding the failure, it was decided to open the bottom outlet gate in the dam. This would permit a controlled release of water and prevent damage to a highway bridge under construction downstream of the dam. The response of the dam had been monitored intermittently during filling by survey measurements made on targets located on the downstream face of the dam.

Within a few hours of the bottom outlet gate being opened on December 2, 1959, the dam failed in a single movement without warning. The city of Frejus, 7 km (4.5 miles) downstream of the damage suffered heavy losses resulting from the release of water and debris. More than 300 people died in the disaster. Blocks of concrete from the dam were washed as much as 1.5 km (1 mile) downstream.

The failure of Malpasset Dam represented the first failure of an arch dam. The suddenness of the failure, given that nothing abnormal had been detected at the dam within the hours preceding the event, added to the intrigue.

Lessons Learned:

A number of valuable lessons that changed certain design and construction methods for future dams resulted from the Malpasset Dam failure. Most notably, design and construction methods to mitigate the effects of uplift pressures in the foundation of the dam were developed. Further, the use of appropriately located and selected monitoring instruments increased following the dam failure.

References:

Engineering News-Record, (1959), "French Dam Collapse: Rock Shift was Probable Cause", December 10, pp. 24-25.

Bellier, J., (1967), "Le Barrage de Malpasset", Traveuse, July.

Bellier, J., and Londe, P., (1976), "The Malpasset Dam", Proceedings of Engineering Foundation Conference on the Evaluation of Dam Safety, California.

Londe, P., (1987), "The Malpasset Dam Failure", Engineering Geology, Vol. 24, pp. 295-329.

VAJONT DAM - (1963)

Construction of a 276 m (900 ft) high double-arched dam across the Vajont River valley in Northern Italy was undertaken between 1957 and 1960. The dam created a reservoir with an estimated full capacity of about 169 million cu m (220 million cu yds). In 1959, concerns about potential slope stability problems in the reservoir were raised and resulted in further analyses being undertaken. These studies confirmed that there was a slide problem, however there was disagreement as to the volume of material which would be involved in a slide. Predictions ranged from relatively small volumes associated with local surficial movements 10 to 20 m (33 to 66 ft) deep up to volumes of the magnitude of the actual slide as a result of deep-seated movements.

Recognition of potential slope stability problems resulted in the installation of a monitoring program in 1960 and a staged reservoir filling schedule. It was considered that the rate of movements could be controlled by raising and lowering the reservoir water level. Results from the monitoring program over the next 3 years, which relied primarily on devices to detect surface movements, did confirm a relationship between the reservoir level and the slide mass movement however it failed to give a warning of the rate at which the failure ultimately occurred.

On October 9, 1963 the southern rock slope of the reservoir failed over an approximately 2 km (1.2 miles) length. Movement rates of the slide mass of approximately 275 million cu m (360 million cu yds) during the failure were estimated to be of the order of 25 m per sec (80 ft per sec) as opposed to the typical rates recorded during the previous three years of monitoring of less than 1 cm per day (0.4 in. per day) to at most 20 cm per day (8 in per day) on the day of the failure. The massive slide mass came to rest some 360 m (1180 ft) laterally and 140 m (429 ft) upward on the opposite bank of the reservoir. The top of the slide mass was some 160 m (525 ft) above the crest of the arch dam. At the time of the failure, the reservoir was about 66% full and contained an estimated 115 million cu m (150 million cu yds) of water. The water level had been lowered in a controlled fashion by approximately 10 m (33 ft) in the preceding two weeks. As the slide mass plunged into the reservoir, the water was displaced over the dam crest in a stream estimated to be up to 245 m (800 ft) above crest level. Five villages including 2040 lives were lost.

Lessons Learned:

The massive slide into the Vajont reservoir yielded a number of important lessons with regard to the analysis and monitoring of slope movements. The difficulty of predicting when a slide mass will accelerate or fail became evident and the difficulty of estimating changes in states of stress and strength during sliding was reinforced.

References:
Kiersch, G.A., (1964), "Vajont Reservoir Disaster", Civil Engineering, pp. 32-40.

Muller, L., (1964), "The Rock Slide in the Vajont Valley", Rock Mechanics and Engineering Geology, Vol. II, pp. 148-212.

Muller, L., (1968), "New Considerations in the Vajont Slide", Rock Mechanics and Engineering Geology, Vol. VI, pp. 1-91.

Muller, L., (1987), "The Vajont Catastrophe - A Personal Review", Engineering Geology, Vol. 24, pp. 423-444.

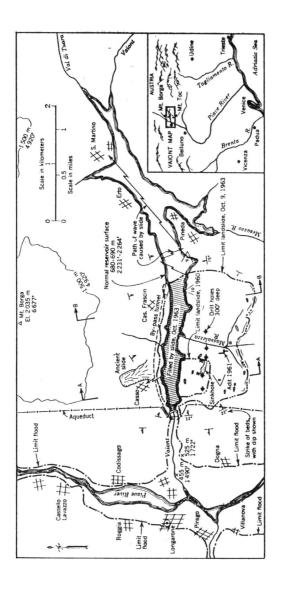

Map of Vajont Dam Area Showing Limits of Slide and Resulting Wave
(Kiersch, 1964; Reprinted by permission of ASCE)

LOWER SAN FERNANDO DAM - (1971)

Construction of a 40 m (130 ft) high hydraulic-fill earth dam was initiated in 1912 as part of a reservoir system in San Fernando, California. Hydraulic fill was placed between 1912 and 1915. The material was excavated from the bottom of the reservoir and discharged through sluice pipes located at starter dikes on the upstream and downstream edges of the dam. This construction configuration resulted in upstream and downstream shells of sands and silts and a central core region of silty clays. A stratum of variable thickness between 3 and 5 m (10 and 16 ft) of reworked weathered shale (silty sand to sand size) was placed on top of the hydraulic fill material in 1916. Several additional layers of roller compacted fill were placed between 1916 and 1930 and raised the dam to its final height of about 40 m (130 ft). A roller compacted berm was placed on the downstream side in 1940. The dam created a reservoir with an estimated full capacity of about 25 million cu m (33 million cu yds).

On February 9, 1971, the San Fernando earthquake occurred with an estimated Richter magnitude 6.6. At the time of the earthquake, the water level in the reservoir was about 11 m (36 ft) below the crest. This reduced level was in part the result of an earlier seismic stability analysis which imposed a minimum operating freeboard criteria of 6 m (20 ft). During and immediately after the earthquake, a major slide involving the upstream slope and the upper part of the downstream slope occurred. As a result of the slide, a freeboard of about 1.5 m (5 ft) remained. Given the likelihood of further damage in the presence of after-shocks, 80,000 people living downstream of the dam were evacuated until the water level was lowered to a safe elevation over a 4 day period.

Lessons Learned:

The near-disastrous slide of the San Fernando dam had a major impact on future earth-dam design and construction procedures. The availability of site specific seismograph data permitted extensive analyses to be performed and showed among other factors, the potential problems with hydraulic fill structures and. the need for revised procedures for dynamic stability analyses of earth dams.

References:

Seed, H.B., Lee, K.L., Idriss, I.M., and Makdisi, F.I., (1975), "The Slides in the San Fernando Dams during the Earthquake of February 9, 1971", ASCE, Journal of Geotechnical Engineering Division, Vol. 101, No. GT7, pp. 651-688.

Seed, H.B., Idriss, I.M., Lee, K.L., and Makdisi, F.I., (1975), "Dynamic Analysis of the Slide in the Lower San Fernando Dam during the Earthquake of February 9, 1971", ASCE, Journal of Geotechnical Engineering Division, Vol. 101, No. GT9, pp. 889-911.

Castro, G., Seed, R.B., Keller, T.O., and Seed, H.B., (1992), "Steady State Strength Analysis of Lower San Fernando Dam Slide", ASCE, Journal of Geotechnical Engineering Division, Vol. 118, No. GT3, pp. 406-427.

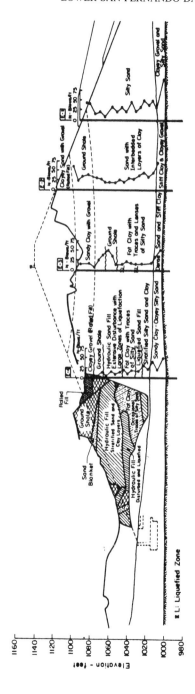

Post-Earthquake Section through Lower San Fernando Dam
(Seed et al., 1975; Reprinted by permission of ASCE)

TETON DAM - (1976)

A 90 m (300 ft) high zone-filled earth dam was constructed in a steep walled canyon eroded by the Teton River in Idaho. The typical cross-section consisted of a wide silt core with upstream and downstream shells consisting mainly of sand, gravel and cobbles. In the main section of the dam, the impervious core was keyed into the foundation alluvium some 30 m (100 ft) to serve as a cut-off trench. Lesser cut-off trenches were excavated at both abutments through the permeable rock. Reservoir filling commenced in November, 1975 at an intended rate of about 0.3 m (1 ft) per day. Delays in completing the construction of outlet works combined with heavier than expected spring melt run-off resulted in a filling rate up to 1.2 m (4 ft) per day in May, 1976.

The dam failed on June 5, 1976 when the water level in the reservoir was at an elevation 9 m (30 ft) below the embankment crest and just 1 m (3 ft) below the spillway crest. Breaching of the dam crest and complete failure was preceded over a period of two days by increasing quantities of seepage being observed initially some 460 m (1500 ft) downstream and later on the downstream face of the dam. Noticeable increases in the seepage flow rate from the face of the dam adjacent to the abutment about 40 m (130 ft) below the crest occurred during the morning of June 5. By approximately 10:30 am the flow rate had increased to about 0.4 cu m per sec (15 cu ft per sec). This quantity continued to increase as a 1.8 m (6 ft) diameter "tunnel" formed perpendicular to the axis of the dam. By 11:00 am a vortex was observed in the reservoir. The seepage flow rate increased rapidly from this time onwards, accompanied by progressive upward erosion of the "tunnel" crown. The dam crest was breached at about 11:55 am with complete failure of the dam ensuing. The flooding downstream after the failure of the dam resulted in the loss of 14 lives and caused an estimated $400 million in damage.

Lessons Learned:

The failure of Teton Dam was important in that no dam of such a height had previously failed. It provided important lessons relating to the need for instrumentation; the need for protective filters to prevent uncontrolled seepage erosion; the design of cut-off trenches; consideration of the impact of frost action; and the importance of adequate compaction control criteria and methods.

References:

Chadwick, W.L., (1977), "Case Study of Teton Dam and its Failure", Proceedings of 9th International Conference on Soil Mechanics and Foundation Engineering, Case History Volume, Tokyo.

Seed, H.B., and Duncan, J.M., (1981), "The Teton Dam Failure - A Retrospective Review", Proceedings of 10th International Conference on Soil Mechanics and Foundation Engineering, Stockholm, Vol. 4, pp. 219-238.

Leonards, G.A., and Davidson, L.W., (1984), "Reconsideration of Failure Initiating Mechanisms for Teton Dam", Proceedings of 1st International Conference on Case Histories in Geotechnical Engineering, St. Louis, Vol. 3, pp. 1103-1113.

Seed, H.B., and Duncan, J.M., (1987), "The Failure of Teton Dam", Engineering Geology, Vol. 24, pp. 173-205.

RISSA NORWAY LANDSLIDE - (1978)

A landslide including seven farms in an area of 330,000 sq m (0.127 sq miles) occurred in Rissa, Norway, on April 29, 1978. Rissa is located just north of the city of Trondheim, Central Norway. The Trondheim and Oslo regions are the two large marine clay areas of Norway. The Rissa event involved 5 to 6 million cu m (7 to 8 million cu yds) of slide debris and is the biggest to occur in Norway in this century. Of forty people caught within the slide area, only one was killed. Many people thus witnessed the event which was recorded on amateur films as it occurred. The slide is thus exceptionally well documented.

The landslide was due to complete liquefaction of quick clay and took place retrogressively in large sections triggered by the excavation and stockpiling of 700 cu m (900 cu yds) of soil placed by the shore of Lake Botnen. Excavation work for the construction of a new wing to an existing barn was conducted over a two day period. The surplus soil had been placed down by the lake shoreline to extend the land area of the farm. When the earthwork was completed, 70 to 90 m of the shoreline suddenly slid out into the lake. The sliding then developed retrogressively landward from that point. The total duration of the retrogressive sliding occurred over a period of about 6 minutes.

Lessons Learned:

The Rissa landslide is an example of a situation where "tampering with nature" can lead to catastrophic results. Construction in quick clay areas must be preceded by careful study and analysis.

References:

Grande, L.,(1978), "Description of the Rissa Landslide, 29 April 1978, Based on Eyewitness Accounts," Tech. Univ. of Norway, Trondheim, Inst. of Soil Mech. and Found. Engrg.

Gregersen, O., (1981), "The Quick Clay Landslide in Rissa, Norway. The Sliding Process and Discussion of Failure Modes," NGI Publication No. 135, Norwegian Geotechnical Institute, pp. 1-6.

NERLERK BERM FAILURE - (1983)

Construction of an underwater sand berm designed as part of an offshore artificial sand island structure for hydrocarbon exploration at Nerlerk in the Canadian Beaufort Sea was undertaken during the 1982 and 1983 open-water construction seasons. A series of slides occurred during the 1983 construction season which ultimately lead to abandonment of the berm.

The artificial island was to be constructed at a site where the water was about 45 m (150 ft) deep. The design called for a sand berm with sideslopes of the order of 5H:1V extending from the seafloor to the design berm top elevation of about 10 m (33 ft) below sea level. A steel caisson superstructure would then be ballasted on top of the berm to permit hydrocarbon exploration. About 1.3 million cu m (1.7 million cu yds) of dredge material was placed during the 1982 construction season using bottom dump hopper dredges which hauled the medium sand fill from the Ukalerk borrow area. An additional 1.8 million cu m (2.3 million cu yds) of clean medium sand material was dredged during the 1982 construction season from a nearby borrow area and was pumped via floating pipeline to a discharge barge situated at the berm site.

Construction recommenced in the summer of 1983 with the deep suction dredger to discharge a barge floating pipeline system. Bathymetric surveys revealed that a significant part of the berm disappeared on July 20, 1983. Several additional construction slides were noted in the following weeks as efforts were made to complete the berm to the design elevation and assess the cause of the initial failure. Construction at the initial site was finally abandoned in early August. Hydrocarbon exploration activities were ultimately conducted using an alternative drilling system. Nevertheless, berm construction cost losses in excess of $50 million were incurred.

Lessons Learned:

A number of very valuable lessons were learned from this case history with respect to the need to carefully control dredge construction methods. Post failure investigations indicated fill materials in certain zones was in a very loose state. Conflicting triggering mechanisms have been postulated.

References:

Sladen, J.A., D'Hollander, R.D., Krahn, J., and Mitchell, D.E., (1985), "Back Analysis of the Nerlerk Berm Liquefaction Slides", Canadian Geotechnical Journal, Vol. 22, No. 4, pp. 579-579-588.

Sladen, J.A., D'Hollander, R.D., Krahn, J., and Mitchell, D.E., (1987), Closure to "Back Analysis of the Nerlerk Berm Liquefaction Slides", Canadian Geotechnical Journal, Vol. 24, No. 1, pp. 179- 185.

Been, K., Conlin, B.H., Crooks, J., Fitzpatrick, S.W., Jefferies, M.G., Rogers, B.T., and Shinde, S., (1987), Discussion of "Back Analysis of the Nerlerk Berm Liquefaction Slides", Canadian Geotechnical Journal, Vol. 24, No. 1, pp. 170-179.

Konrad, J.M., (1991), "The Nerlerk Berm Case History: Some Considerations for the Design of Hydraulic Sand Fills", Canadian Geotechnical Journal, Vol. 28, No. 4, pp. 601-612.

CARSINGTON EMBANKMENT - (1984)

A 1250 m (410 ft) long, 30 m (100 ft) high zone filled earth embankment was being constructed as part of a water storage scheme for the Severn-Trent Water Authority to regulate flows in the River Derwent in England. The scheme was designed so that water would be diverted through a 10 km (6 mile) long tunnel during the winter and stored in the reservoir which had an estimated capacity of 35 million cu m (46 million cu yds). Water would be released from the reservoir when the water level in the river was low.

The central clay core of the embankment connected to a shallow trench that was excavated upstream of the centerline into the weathered gray foundation mud stone. A grout curtain extended below the base of the trench. Fill in the upstream and downstream shells was classified as Type I and Type II. The Type I fill used immediately upstream and downstream of the clay core was described as a yellow/brown mottled clay with mud stone peas < 5 mm (0.2 in.) and pebbles. The Type II soil which was located in the outer portions of the shells was to be the same general type of material but without pebbles.

Construction of the embankment began in July, 1982 and reached a height of about 6 m (20 ft) at the end of the construction in late October. Construction during the second year between April and October added an additional 15 m (50 ft) to the embankment. Placement of an additional 4 m (13 ft) of fill took place in the two months preceding the failure. Tension cracks over an approximate 65 m (213 ft) length were first observed on the crest of the embankment on June 4, 1984 when about 1 m (3 ft) of crest remained to be placed. Thirty six hours later, a 400 m (1300 ft) section of the upstream slope failed with a maximum horizontal displacement of 15 m (50 ft).

Lessons Learned:

The failure of Carsington Dam is considered of importance in that it led to additional attention being given to the role of the construction equipment and procedures in the subsequent stability of a structure. In this case, the compaction equipment selected and the rate of fill placement are considered to have been key factors in the observed failure. In addition, the importance of selecting instrumentation which can provide a precursor to a failure was reinforced.

References:

New Civil Engineer, (1984), "Weak Ground Cited as Carsington Fails", June 14.

Skempton, A.W., (1985), "Geotechnical Aspects of the Carsington Dam Failure", Proceedings of 11th International Conference on Soil Mechanics and Foundation Engineering, San Francisco, Vol. 5, pp. 2581-2591.

Rowe, P.W., (1991), "A Reassessment of the Causes of the Carsington Embankment Failure", Geotechnique, Vol. 41, No. 3, pp. 395-421.

GEOENVIRONMENTAL FAILURES

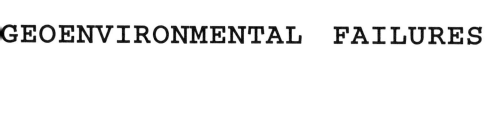

LOVE CANAL - (1978)

In the late 1890s, an entrepreneur named William T. Love initiated the construction of a canal linked to the Niagara River that was intended to be used as a source for hydroelectric power and attract industry to the Niagara Falls, New York area. The project was abandoned when a section of canal about 1000 m (3200 ft) long and 24 m (80 feet) wide had been excavated to a depth of the order of 6 m (20 ft). In 1942, the Hooker Chemical and Plastic Company purchased the abandoned excavation site from the Niagara Power and Development Company and began using the canal excavation as a dumpsite for industrial wastes that included pesticide residues, process slurries and waste solvents. In total, approximately 22,000 tons of waste contained in metal drums was placed in the excavation during an eleven year period. Later studies would show that more than 200 different chemical compounds including at least 12 known carcinogens were present. Once filled, the excavation was capped with a loose soil cover.

The Hooker Chemical and Plastic Company sold the land to the City of Niagara Falls for $1 in 1953. A residential sub-division and school were subsequently built on the site. During the mid 1970s, a number of the area residents began developing a variety of illnesses. In 1975 and 1976, significant precipitation raised the groundwater table and caused portions of the land to subside. Waste drums appeared at the surface and chemical odors were very noticeable near sewer manholes. Some of the houses nearest the old canal site had basements where black sludge that produced a strong smell was observed oozing through the walls. After several preliminary studies revealed increased rates of birth defects and miscarriages among area residents, the government declared a state of emergency and two separate evacuations took place in 1978 (area immediately adjacent to canal involving approximately 240 families) and in 1980 (larger area involving approximately an additional 570 families). Extensive and protracted clean-up at the site including removal of about 5,700 cu m (7,500 cu yds) of contaminated soil, placement of a 1 m (3 ft) clay cover over some 7 ha (18 acres) immediately above and adjacent to the canal, placement of a second composite HDPE clay liner system over some 16 ha (40 acres) and cleaning of about 200 m (5,000 ft) of sewer has resulted in estimated costs in excess of half a billion dollars to date and there remains distinct debate about the future habitability of the area.

Lessons Learned:
The Love Canal site was one of the landmark environmental failures which came to light in the 1970s and led to the passing of the Comprehensive Environmental Response, Compensation and Liability Act in 1980. As such, it clearly changed the attention given to environmental assessment prior to real estate transactions and the assignation of responsibility for inappropriate disposal of hazardous wastes.

References:
Brown, Michael H., (1979), "Love Canal and the Poisoning of America", The Atlantic, December, pp. 33-47.

U.S. EPA, Environmental Monitoring at Love Canal, (1982), EPA 600/4-82-0309.

U.S. EPA, Superfund Record of Decision: Love Canal, 93rd Street, (1988), NY, September, EPA ROD/R02-88/063.

TIME, (1988), "Welcome Back to Love Canal", October 10, Vol. 132, pp. 49.

VALLEY OF THE DRUMS - (1978)

The Valley of the Drums is located in Northern Bullitt County, Kentucky near the town of Brooks on land that was owned by Mr. A.L. Taylor prior to his death in 1977. The site, which was first identified as a waste disposal facility by the Kentucky Department of Natural Resources and Environmental Protection (KDNREP) in 1967, consisted of approximately 5 ha (13 acres) which was used as an uncontrolled dump site.

More than 27,000 drums of industrial waste were discovered on the site in 1978 by KDNREP officials. Later studies however, estimated that more than one hundred thousand drums had been delivered to the site. Due to space limitations, the contents of many of the drums were dumped in open pits and trenches that were excavated on site. The empty drums were then either sold or crushed. The trenches were subsequently covered with on-site soil fill.

Over a period of time, the conditions of many of the drums on site deteriorated and the contents spilled onto the ground and were flushed into a nearby creek by storm water runoff. Additional contaminants seeped into the creek from the disposal trenches. Frequent complaints about strong odors along the creek bed were received from adjacent property owners. In March, 1979, unusually large amounts of wastes were flushed into the creek by melting snow runoff and resulted in an emergency response by the Environmental Protection Agency. Subsequent analysis of water and soil samples indicated substantial levels of contamination by heavy metals and polychlorinated biphenyl's along with some 140 other chemical substances. Extensive remedial actions were undertaken in 1986 and 1987 to control surface run-off and reduce the future impact of buried wastes.

Lessons Learned:

The Valley of the Drums site was one of the landmark environmental failures which came to light in the 1970s and led to the passing of the Comprehensive Environmental Response, Compensation and Liability Act in 1980. As such, it clearly changed the attention given to facility permitting and compliance monitoring and the assignation of responsibility for inappropriate disposal of hazardous wastes.

References:

USEPA, (1980),Valley of the Drums, Bullitt County, Kentucky. Oil and Special Materials Control Division, Washington, D.C., Publication No. EPA-430/9-80-014, 53 pp.

Metcalf & Eddy, Inc., (1984), Feasibility Study Addendum and Endangerment Assessment: A.L. Taylor Site, EPA Contract #68-01-6769, September.

USEPA, (1986), Record of Decision, Remedial Alternative Selection, A.L. Taylor Site, Brooks, Kentucky. Region IV Administrative Records Division, Atlanta, Ga., Document No. 001422, 109 pp.

Resource Applications, Inc., (1992), A.L. Taylor Five-Year Review Final Report, EPA Contract #68-W9-0029, June.

STRINGFELLOW ACID PITS - (1980)

Between 1956 and 1972, the Stringfellow Quarry Company operated a state authorized hazardous waste disposal facility 8 km (5 miles) Northwest of Riverside, California. Approximately 155 million liters (34 million gallons) of industrial wastes from metal finishing, electroplating and DDT production industries were placed in unlined evaporation ponds. The wastes disposed of in these ponds migrated into the underlying highly permeable soils into the groundwater and resulted in a contaminated plume which extended some 3 km (2 miles) downstream.

The site was voluntarily closed in 1972. The California Regional Water Quality Control Board declared the site a problem area. Between 1975 and 1980, approximately 30 million liters (6.5 million gallons) of liquid wastes and DDT contaminated materials were removed and a policy to contain the waste and minimize further contaminant migration was adopted. The Environmental Protection Agency (EPA) led an additional clean-up effort in 1980 which resulted in an additional 45 million liters (10 million gallons) of contaminated water being removed from the site. In 1983, the site was added to the National Priorities List as California's worst environmental hazard. Since that time, 4 Records of Decision have been issued by the EPA outlining required clean-up measures. Estimates of clean-up costs as high as three-quarters of a billion dollars have been cited along with indications that the work could last into the next century.

Lessons Learned:

The Stringfellow Acid Pits site was one of the landmark environmental failures which came to light in the 1970s and led to the passing of the Comprehensive Environmental Response, Compensation and Liability Act in 1980. As such, it clearly changed the attention given to environmental assessment prior to real estate transactions and the assignation of responsibility for inappropriate disposal of hazardous wastes. One interesting fact about this site is that it is one of the few cases to date where a government has been found liable for contributing to an environmental problem. In this case, the State of California was found at fault as a result of its actions in selecting this particular site for disposal and subsequently controlling activities there.

References:

U.S. Environmental Protection Agency, (1983), EPA/ROD/R09-83/005 Stringfellow Record of Decision: Stringfellow Acid Pits Site; Initial Remediation Measure, July.

U.S. Environmental Protection Agency, (1984), EPA/ROD/R09-84/007 Stringfellow Record of Decision: Stringfellow Acid Pits: Untitled, July.

U.S. Environmental Protection Agency, (1987), EPA/ROD/R09-87/016 Stringfellow Record of Decision: Stringfellow Hazardous Waste Site: 2nd Remedial Action, July.

U.S. Environmental Protection Agency, (1990), EPA/ROD-R09-90/048 Stringfellow Record of Decision: Stringfellow Hazardous Waste Site: 4th Remedial Action.

Engineering News Record, (1992), "EPA, PRP's Sign Pact to Clean Stringfellow", August 17, Vol. 229, pp. 14.

New York Times, (1993), "Largest-ever Toxic-waste Suit Opens in California", February 5, Vol. 142.

SEYMOUR RECYCLING FACILITY - (1980)

The Seymour Recycling Corporation operated a 5.6 ha (14 acre) site as a recycling center to process industrial waste chemicals approximately 3 km (2 miles) Southwest of the city of Seymour in Jackson County, Indiana. From 1970 to 1980, wastes were collected in drums and bulk storage tanks. By 1980, there were approximately 100 storage tanks and about 50,000 drums on site. Many of the drums were in poor condition and their contents had leaked while others had no lids. Widespread contamination of the underlying soil and groundwater occurred and resulted in on-site fires and unpleasant toxic odors being reported by neighboring residents.

As a result of a fire in 1980, chemical runoff from the site posed a threat which resulted in approximately 300 people being temporarily evacuated. The facility was subsequently closed. Over the next 4 years, the majority of tanks and drums were removed by those identified as being potentially responsible parties.

A range of remedial measures were implemented. An embankment was constructed around the site to control surface runoff. Approximately 900,000 liters (200,000 gallons) of flammable chemicals were incinerated. More than 450,000 liters (100,000 gallons) of inert liquids were injected into a deep well. Other wastes and containers including more than 23,000 cu m (30,000 cu yds) of drums, sludges and contaminated soil were placed in a hazardous waste landfill. Despite these remedial measures, monitoring wells showed that a contaminated groundwater plume extended more than 120 m (400 ft) off site by 1985. Tests indicated the presence of heavy metals and numerous organic and phenols within the soil and groundwater. The site was placed on the National Priorities List as a result of being designated the most serious environmental threat in Indiana. Extensive long term remedial activities are currently in progress and include on-site incineration of some contaminated soils, use of bioremediation technology to assist in the cleanup, the installation of a vapor extraction system to remove volatile organic compounds from the vadose zone and the installation of a pump and treat system to stabilize the contaminated groundwater plume and treat it at the Seymour waste water treatment plant.

Lessons Learned:

The Seymour Recycling Facility site was one of the landmark environmental failures which occurred in the 1970s and contributed to the passing of the Comprehensive Environmental Response, Compensation and Liability Act in 1980. As such, it clearly changed the attention given to facility permitting and compliance monitoring and the assignation of responsibility for inappropriate disposal of hazardous wastes.

References:

US EPA, (1984), "Hazardous Waste Sites, Description of Sites on Current National Priorities List, October 1984", EPA/HW 8.5, December.

US EPA, (1986), "Superfund Record of Decision: Seymour, IN", EPA/ROD/R05-86/046, September.

US EPA, (1987), "Superfund Record of Division: Seymour, IN", EPA/ROD/R05-87/050, September.

KETTLEMAN HILLS WASTE LANDFILL - (1988)

Prior to 1987, construction of a 15 ha (36 acre) landfill known as Unit B-19, was begun with the excavation of a 30 m (100 ft) deep oval-shaped "bowl" for a Class I hazardous-waste treatment-and-storage facility at Kettleman City, California. The "bowl" was designed to have a nearly horizontal base with side slopes of 1:2 or 1:3 into which waste fill was to be placed. A multilayer liner system consisting of impervious geomembranes, clay layers, and drainage layers, line the base and sides of the repository to prevent the escape of hazardous materials (leachates) into the underlying and surrounding ground and underlying ground water.

The lining of 6 ha (15 acres) (Phase I-A) of the northern end of the "bowl" was completed first and placement of solid hazardous waste was begun in early 1987.

On March 19, 1988, after the waste pile reached a maximum height of about 27 m (90 ft) in Phase I-A with no prior indication of distress, a slope stability failure occurred with lateral displacements of the waste fill of up to 11 m (35 ft) and vertical settlement of the surface of the fill of up to 4 m (14 ft). Surface cracks, tears and displacements of the exposed portions of the liner system were also visible. It was found that failure had developed by sliding along interfaces within the composite, multilayered geosynthetic compacted clay liner system beneath the waste fill.

Because of the fear that the liner system may have also been breached at the facility, a major investigation was undertaken to determine (1) the cause of the failure and (2) appropriate methods of testing and analysis to preclude the possibility of similar failures at other facilities. Comprehensive laboratory tests were performed to evaluate the frictional resistance between contact surfaces of various geosynthetics, geonet, and geotextile; and between these materials and compacted clay liner. The frictional resistance at the interface between these materials are characterized as being insufficient to maintain stability of the waste pile.

Lessons Learned:

The interface frictional resistance of liner materials are affected by various properties. These properties include degree of polishing, whether the interfaces are wet or dry, and in some cases, the relative orientation of the layers to the direction of shear stress application. Some small variations in properties also exist between one batch and another of the geosynthetic materials. Wetting of the compacted clay liner at its contact with the geosynthetics also affects the interface frictional resistance.

The test and observed field results also indicate the desirability of performing similar test programs for proposed new facilities to establish design parameters, until such a time as more data and experience are available.

References:

Mitchell, J.K., Seed, R.B. and Seed, H.B., (1990), "Kettleman Hills Waste Landfill Slope Failure. I: Liner-System Properties", Journal of Geotechnical Engineering, ASCE, Vol. 116, No. 4, April, pp. 647-668.

Seed, R.B., Mitchell, J.K. and Seed, H.B., (1990), "Kettleman Hills Waste Landfill Slope Failure. II: Stability Analysis", Journal of Geotechnical Engineering, ASCE, Vol. 116, No. 4, April, pp. 669-690.

Martin, J.P., Koerner, R.M. and Whitty, J.E., (1984), "Experimental Friction Evaluation of Slippage between Geomembranes, Geotextiles and Soils," Proceedings of International Conference on Geomembranes, Denver, Colorado, pp. 191-196.

Mitchell, J.K., Seed, R.B. and Chang, M., (1993), "The Kettleman Hills Landfill Failure: A Retrospective View of the Failure Investigations and Lessons Learned," Proc. Third International Conference on Case Histories in Geotechnical Engineering, St. Louis, Vol. II, pp. 1379-1392.

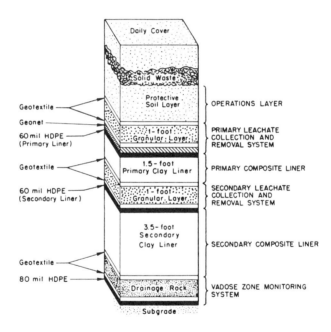

Schematic Illustration of Multilayer Liner System at Base of Landfill
(Seed et al., 1980; Reprinted by permission of ASCE)

BRIDGE FAILURES

THE ASHTABULA BRIDGE - (1876)

The Ashtabula Bridge was constructed during 1863-1865 over a stream in Ashtabula, Ohio. The bridge collapsed on December 29, 1876 after eleven years of service.

The conceptual design of the Ashtabula bridge was carried out by Amasa Stone, the president of the Cleveland, Painesville and Ashtabula Railroad, which later became a part of the Lake Shore and Michigan Southern Railway. The bridge was a simply-supported, parallel chord, Howe truss with a span of 47 m (154 ft) consisting of 14 panels of 3.4 m (11 ft) each. The depth of the truss was 6 m (20 ft) and the center-to-center spacing of the trusses was 5 m (17 ft). The bridge was a deck structure carrying double track railway. Stone employed Joseph Tomlinson to determine the member sizes and prepare the fabrication drawings. Tomlinson carried out the task assigned to him and supervised the entire fabrication. However, he did not participate in the supervision of the construction.

The diagonals subjected to compressive forces consisted of several iron I-beams in parallel. The top chord consisted of segments, two panels long, fitting between the lugs on the iron casting. The beams carrying the timber floor rested directly on the top chord, thus introducing flexure in the top chord of the truss. The position of each track was such that the live load of one train was carried predominantly by the truss near to that track.

Failure of the bridge took place at about 7:30 p.m. on December 29, 1876 during a severe snowstorm. A train with two locomotives was crossing the bridge heading west at an estimated speed from 20 to 25 km per hour (12 to 15 miles per hour). As the first locomotive was about to complete the crossing, the bridge began to fail. The second locomotive, its tender, and eleven cars fell 20 m (65 ft) into Ashtabula Creek. The accident was a national tragedy causing some 80 deaths.

On January 12, 1877 the Legislature of Ohio appointed a joint committee to investigate the causes of failure and report to the legislature on January 30, 1877. In addition, the coroner's jury appointed an engineer to investigate the causes of the disaster. A third major investigation of the failure was carried out by Charles MacDonald, probably on behalf of the ASCE.

It was determined that the failure occurred in the second and third panels of the south truss, though the investigators were undecided whether it was the top chord, or the compressive brace that initiated the failure. The Joint Committee concluded the bridge failed because of inadequate inspection. The coroner's jury made the same point, adding that iron bridges were in their infancy and that an experiment should not have been made on a bridge with such a deep chasm. Charles MacDonald suggested that a fatigue crack originated at the flaw in the lug and propagated under repeated stress cycles.

Lessons Learned:

The failure bolstered the call for consulting bridge engineers and standard design specifications. It brought in focus the question of reliability of iron castings.

References:

"Report of the Joint Committee Concerning the Ashtabula Bridge Disaster", (1877), Nevins Myers State Printers, Columbus, Ohio.

McDonald, C., (1877), "The Failure of Ashtabula Bridge", Transactions, ASCE, Vol. 6, pp. 74-87.

Gasparini, D.A. and Fields, M, (1993), "Collapse of Ashtabula Bridge on December 29, 1876", Journal of Performance of Constructed Facilities, ASCE, Vol. 7, No. 2, pp. 109-125.

THE TAY BRIDGE - (1879)

Spanning the Firth of Tay at Dundee, Scotland, the Tay Bridge was opened to railway traffic on May 31, 1878. The bridge collapsed on Sunday, December 28, 1879. With a length of 3146 m (10,321 ft), when completed it was the longest bridge in the world ever built over a stream. The bridge was made of eighty-five simply-supported spans as follows: six at 8.2 m (27 ft), fourteen at 20.6 m (67.5 ft), fourteen at 21.5 m (70.5 ft), two at 26.8 m (88 ft), twenty-one at 39.3 m (129 ft), thirteen at 44.5 m (146 ft), one at 49.4 m (162 ft), one at 51.8 m (170 ft) and thirteen at 74.7 m (245 ft). The shorter spans were deck trusses, the longer ones were through trusses and the 51.8 m (170 ft) span section was a bowstring truss.

Many difficulties were encountered during the construction. The Tay, being a tidal river, was subjected to floods and exposed to wind gusts that constantly hampered the construction operations. Many changes were introduced as the construction progressed and the final structure differed in many ways from what was originally planned. The modifications were not properly recorded.

The bridge failed on Sunday, December 28, 1879 during a major storm at the estuary of the Tay. A train, traveling north from Edinburgh to Dundee, had to cross the bridge just before reaching Dundee but never reached its destination. It is not known whether the bridge collapsed over a portion of its length under the train as it was passing over, or whether it had already fallen before the train reached it. Thirteen spans of the bridge fell over. The accident resulted in the death of 75 people.

Lesson Learned:

The principal cause of the failure was improper estimate of the wind force. According to the records the wind force was taken to be 0.5 kPa (10 psf) which should have been in the order of 2.4 kPa (50 psf). There were also some shortcomings in inspection, particularly at the Wormit foundry where the bridge elements were being manufactured. The Court of Inquiry determined that Thomas Bouch, the engineer in charge, was entirely responsible for the faults in design and principally responsible for faults in construction and maintenance.

References:

"The Tay Bridge Accident", (1880), Engineering, Vol. 20, Jan. 2, pp. 11-13; Jan. 9, pp. 31-34; Jan. 30, pp. 92-94; Feb. 13, pp. 132-133; Mar. 5, pp. 191-193.

"The Tay Bridge", (1880), Engineering News, Vol. 7, Jan. 10, pp. 10-13; Jan. 24, pp. 36-39; Jan. 31, pp. 44-46, 46-47; Feb. 14, pp. 59-60; Feb. 28, pp. 79-84; Mar. 20., pp. 107; Apr. 24, pp. 146; July 31, pp. 259-260; Aug. 21, pp. 284.

"Tay Bridge Inquiry", (1880), Engineering, Vol. 20, Apr. 23, pp. 320-323; Apr. 30, pp. 335-339; May 7, pp. 363-366; May 14, pp. 387-390.

"The Tay Bridge ", (1880), Engineering, Vol. 21, July 9, pp. 24-25; July 30, pp. 96.

"Historic Accidents and Disasters, The Tay Bridge", (1941), The Engineer, Vol. 172, July 11, pp. 18-20; July 18, pp. 34-35.

Prebble, J., "Disaster at Dundee", (1956), Harcourt Brace & Company, New York.

THE QUEBEC BRIDGE - (1907 & 1916)

It took some twenty years to design and build the Quebec Bridge over the St. Lawrence River and during its construction there were two accidents. The first accident occurred on August 29, 1907, fifteen minutes before the end of the working day when the south anchor arm of the bridge collapsed. Seventy-four workers were killed. The second accident took place on September 11, 1916 when, during the lifting of the center span of the newly reconstructed bridge, it fell in the water. Another eleven deaths occurred in this accident. The bridge was opened to traffic in August 1919. The bridge was a double-track railway cantilever truss structure with clear span of 1,800 ft. It was at the time the longest span ever attempted anywhere.

The cause of the first failure was improper design of the compression members of the south anchor arm. In this instance, there was warning prior to the failure. The compression member in the second shoreward panel from the south pier of the anchor arm showed distortion of 5.7 cm (2.25 in.) two days prior to failure.

The second failure occurred when the simple span weighing 50 MN (5000 tons) was being lifted in place. The cruciform casting of the southwest corner of the span failed and allowed the corner of the span to drop on the lifting girder. The blow kicked the girder backward with the result that the span was left hanging unevenly on three corners only and it failed.

Lessons Learned:

The immediate cause of the first failure was inadequate design of compression members which buckled during the construction. However, considering the enormous size of the structure, not sufficient thought was given to the design and the supervision was totally inadequate. The failure was a great tragedy and led to the development of design specifications for bridge structures.

The second failure, which was triggered by the breakage of the rocker shoes, caused the suspended span to fall into the water. Clearly extreme care is necessary to ensure the soundness of the principal elements upon which the lifting of such a heavy load depends.

References:

"The Anchor Pier Towers of the Quebec Bridge", (1907), Engineering Record, Vol. 55, January 12, pp. 34-35.

"The Cause of Quebec Bridge Failure", (1907), Engineering Record, Vol. 56, Sept. 14, pp. 276; Sept. 21, pp. 302.

"Erection Attachments for Bottom Chords and Vertical Posts of the Quebec Bridge", (1907), Engineering Record, Vol. 55, Jan. 19, pp. 71-74.

"Erection of the Main Vertical Posts of Quebec Bridge", (1907), Engineering Record, Vol. 55, Jan. 26, pp. 92-94.

"The Traveller for the Erection of the Quebec Bridge", (1907), Engineering Record, Vol. 55, Feb. 23, pp. 198-210.

"Erection of South Anchor Arm of the Quebec Bridge", (1907), Engineering Record, Vol. 55, Mar. 2, pp. 291-292.

"Erection of the South Cantilever Arm of the Quebec Bridge", (1907), Engineering Record, Vol. 56, Sept. 28, pp. 343-345.

"The Fall of the Quebec Cantilever Bridge", (1907), Engineering News, 58, Sept. 5, pp. 258-265; letters, Sept. 19; pp. 318-319, pp. 319-321.

"Handling Eyebars at the Quebec Bridge", (1907), Engineering Record, Vol. 55, Feb. 9, pp. 153-156.

Horton, H.E., (1907), "Lattice Bars in the Quebec Bridge", Engineering Record, Vol. 56, Sept. 28, pp. 357.

"The Quebec Bridge Disaster", (1907), Engineering, Vol. 84, Sept., pp. 328-330.

"The Quebec Bridge Superstructure Details I", (1907), Engineering Record, Vol. 56, II, July 6, pp. 25-27; III, July 13, pp. 36-37; IV, July 20, pp. 65-66; V, July 27, pp. 89-91; VI, Aug. 3, pp. 130-131; VII, Aug. 10, pp. 159-160; VIII, Aug. 17, pp. 169-170; IX, Aug. 24, pp. 210-211.

"Report of the Royal Commission on Quebec Bridge", (1908), Engineering Record, Vol. 57, April.

Meyers, A.J., (1916), "Quebec Bridge Suspended Span", Engineering News, Sept. 14, pp. 524-529.

Barker, H., (1917), "Quebec Bridge Suspended Span Hung from Cantilever", Engineering News Record, Vol. 79, Sept. 13, pp. 581-588.

"Historic Accidents and Disasters, The First Quebec Bridge", (1941), The Engineer, Vol. 172, Oct. 24, pp. 266-268; Oct. 31, pp. 286-288, pp. 406-409.

Roddis, W.M.K., (1991), "The 1907 Quebec Bridge Collapse: A Case Study in Engineering Ethics", Proceedings, The National Steel Construction Conference, Washington, DC, pp. 23-1 to 23-11.

FALLS VIEW BRIDGE - (1938)

On January 27, 1938 the Falls View arch bridge, just below the Niagara Falls, was torn from its foundation as a result of the worst ice jam on record. The bridge was a tourist attraction, known as the Honeymoon Bridge.

The construction of the bridge started in 1895 and was completed in 1898. The structure was a two-ribbed steel arch of 256 m (840 ft) span. Each rib was a two-hinged truss arch with a uniform depth of 7.9 m (26 ft) and a rise of 45.7 m (150 ft). The chord members were plate and angle box sections. Most of the other members were steel sections with lattice connections. The original wooden deck of 14 m (46 ft) width carried two-track street railway and was supported by unbraced single spandrels. The four concrete and stone foundations rested in solid rock about 12 m (40 ft) above the normal water level.

The arch bridge replaced a suspension bridge which was built in 1868 with wooden towers and floor system. Later the wooden elements of the structure were changed to steel, and on January 10, 1889 the entire deck was carried away by a windstorm. The structure was repaired and opened to traffic, but was judged to be inadequate and it was decided to build an arch bridge.

The ice jam was formed during the night of January 25, 1938 and by the following afternoon it had piled up to a height of 15 m (50 ft) above normal river level, or 3 m (10 ft) above the pins supporting the arch. The ice pack moved downstream like a glacier for about 122 m (400 ft) covering at least 9 m (30 ft) of the upstream truss, causing the failure of many of the bracing members. Shortly thereafter the structure was closed to traffic. The movement of the ice pack was halted, but the upstream truss continued to move very slowly downstream accompanied by further buckling and failure of secondary members. On the afternoon of January 27, the buckled section of the lower chord broke with loud report and the bridge collapsed. The bridge was replaced by the Rainbow Arch, a fixed arch rib of 290 m (950 ft) span.

Lessons Learned:

The principal cause of the failure was the proximity of the ice mass to the structure and the flexibility of the structure. Although it is not always possible to design for unusual natural phenomena, the foundation of bridges should be protected where possible.

References:

"Record Ice Jam at Niagara Falls Wrecks Famous Arch Bridge", (1938), Engineering News Record, Vol. 120, February 3, pp. 161, 168-169.

"Cable and Deck Salvaged from Falls View Bridge", (1938), Engineering News Record, Vol. 120, February 24, pp. 311.

"Falls View Bridge Sinks in River", (1938), Engineering News Record, Vol. 120, pp. 559.

"Ice Power", Engineering News Record, (1938), Vol. 120, pp. 171.

Buck, R.S., (1938), "Niagara Arch Memories," Engineering News Record, 120, February 24, pp. 297-298.

SANDO ARCH - (1939)

On August 31, 1939 during the construction of the reinforced concrete arch bridge over the Angerman River between the villages of Sando and Lunde in Sweden, the timber centering of the forms collapsed. The failure was a major disaster, costing the lives of eighteen construction workers.

The bridge with its span of 264 m (866 ft) was to have become the reinforced concrete arch with the longest span. The arch rib was designed as a three-cell hollow rectangular section having a width of 9.4 m (31 ft) and a depth of 2.7 m (8.75 ft) at the crown and 4.5 m (14.75 ft) at the abutment. The external walls of the hollow section were 30 cm (11-7/8 in.) thick and the internal walls were 20 cm (7-7/8 in.) thick.

The construction schedule called for casting the bottom slab first, followed by the walls and the top slab. The bottom slab was cast in sections and each section was allowed to set before filling the sections in between. At the time of the accident parts of the bottom slab had been completed and the workmen were cleaning the forms in preparation of pouring the center section when the timber centering failed.

The report of the investigating committee appointed after the disaster stated that the failure of the timber arch was caused by the insufficient strength and stiffness of the transverse bracing between the two flanges of the arch. However, later investigations have revealed that the lateral instability of the arch was primarily responsible for the failure. The Sando Arch was rebuilt and opened to traffic August 28, 1943.

Lessons Learned:

The design of timber forms for a structure of that magnitude is a major engineering project. The stability of the timber structure must be assured.

References:

"This Would Have Been World's Longest Concrete Arch", (1939), Engineering News Record, September 28, pp. 34.

Ros, M., (1940), "Zum Lehrgerust-Einsturz der Sando-Brucke Uber den Angermanalv in Schweden", Schweizerische sauzeitung, Vol. 115, January 20, pp. 27-32.

Remarques sur la resistance des cintres en bois de grande ouverture, (1940), L'effondrement du cintre du pont en beton arme sur le fleuve Angerman, en Suede, le Genie Civil, April 27, pp. 283.

Die Sando-Strassenbrucke in Schweden, Schweizerische Bauzeitung, (1943), Vol. 119, July 16, pp. 105.

Granholm, H., (1961), "Sandobrons Bagstallining", Transactions of Chalmers University of Technology, Gothenburg, Sweden, No. 239.

TACOMA NARROWS BRIDGE - (1940)

The Tacoma Narrows suspension bridge, nicknamed "Galloping Gertie", for its propensity to oscillate, opened for traffic on July 1, 1940, and was the most flexible suspension bridge of its time. After being in service for 129 days, failure occurred late in the morning of November 7, 1940. Under a 68 km/h (42 mph) wind, the bridge, which normally vibrated in a vertical plane, began to oscillate with the opposite sides out of phase (torsional mode). The oscillation became extremely violent, until the failure began at mid span, with buckling of the stiffening girders and lateral bracing. The suspenders snapped, and sections of the floor system fell to the water below. Almost the entire suspended span between the towers fell into the water. The side spans, which remained, sagged about 9 m (30 ft) bending back the towers sharply by the pull of the side span cables. The towers, which were fixed at the base by steel anchors deeply embedded in concrete piers, survived but were damaged.

The dramatic collapse of this bridge, captured on film by Professor F. B. Farquharson of the University of Washington, Seattle, who was on the main span just before the failure taking motion pictures of the violent twisting of the bridge deck, provides a spectacular example of the forces developed when dynamic resonance occurs. Because of its excessive motions, the bridge was closed to traffic sometime before the failure occurred. Hence, there was no loss of human life.

The contract drawings consisted of 39 sheets and were signed by the highway and bridge engineers of the State of Washington. The drawings carried the signatures of Moran, Proctor and Freeman, the consultants on substructure and of Leon Moisseiff as consultant of superstructure. In addition, the engineer retained by the Reconstruction Finance Corporation to review the plans commented on the flexibility of the suspended structure, but deferred to the wider experience of Moisseiff. The design was also reviewed and approved by a board of consultants appointed by the Washington Toll Bridge Authority.

The total cost amounted to $6,400,000, of which $2,880,000 was a grant from the Public Work Administration and $3,520,000 was a loan from the Reconstruction Finance Corporation to be repaid by the tolls.

The suspended span was 853 m (2,800 ft) and the two side spans were 335 m (1,100 ft) each. The 853 m (2,800 ft) main span made this the third longest suspension bridge of its time. The roadway was 7.9 m (26 ft) wide and the sidewalks were 1.5 m (5 ft) wide on each side. The spacing between the cables and stiffening girders was 11.9 m (39 ft), a 1:72 span ratio. The main piers were 19.7 m (64.5 ft) by 35.8 m (117.5 ft) in plan. The towers were 128 m (420 ft) in height. Each cable consisted of nineteen strands of 332-No. 6 cold drawn galvanized wires. The diameter of each cable under wrapping was 435 mm (17 1/8 in) with a net area of 1228 sq cm (190 sq in). The original design specified stiffening trusses, but was changed to stiffening girders 2.4 m (8 ft) deep, a 1:350 span ratio.

Based upon previous experience with suspension bridges with shallow stiffening girders, the engineers anticipated oscillations. From the time it was constructed, however, it had a history of disturbing oscillations, up to 1.3 m (50 in) in amplitude, but for the most part they were not considered to be dangerous. Studies of the bridge vibration problem were undertaken, including wind tunnel testing of models at the University of Washington at Seattle, and modifications were made, such as the installation of cable ties attached to concrete anchors. These cables broke three or four weeks before the collapse. Deflector vanes to change the aerodynamic characteristics had also been developed, but their installation was under negotiation at the time of the collapse. After the failure a board on engineers was appointed by the Administrator of the Federal Works Agency to determine the causes of failure.

Lessons Learned:

The Board of Engineers concluded that the bridge was well designed and built to resist safely all static forces. Its failure resulted from excessive oscillations made possible by the extraordinary degree of flexibility of the structure. The Board determined with reasonable certainty that the first failure was the slipping of the cable band on the north side of the bridge to which the center ties were connected. This slipping may have initiated the torsional oscillations. The Board recommended more studies to understand the aerodynamic forces acting on suspension bridges.

Thus, incompetence or neglect were not the cause. The failure was due to the torsional oscillations made possible by the narrow width and small vertical rigidity of the structure. Those actions and forces were previously ignored or deemed to be unimportant in suspension bridge design. This failure emphasized the need to consider aerodynamic effects in the design of a suspension bridge.

References:

"Action of Von Karman Vortices", (1940), Engineering News Record, Vol. 125, pp. 808.

"Aerodynamic Stability of Suspension Bridges", (1940), Engineering News Record, Vol. 125, pp. 670.

"Another Consultant Board Named for Tacoma Span", (1940), Engineering News Record, Vol. 125, pp. 735.

"Board Named to Study Tacoma Bridge Collapse", (1940), Engineering News Record, Vol. 125, pp. 733.

Bowers, N. A. (1940), "Tacoma Narrows Bridge Wrecked by Wind," (1940), Engineering News Record, November 14, pp. 1, 10, 11-12.

Bowers, N. A., (1940), "Model Tests Showed Aerodynamic Instability of Tacoma Narrows Bridge." Engineering News Record, November 21, pp. 40-47.

"Comment and Discussion", (1940), Engineering News Record, Nov. 21, pp. 40.

"Comments and Discussions", (1940), Engineering News Record, Dec. 5, pp. 40-41.

"Dynamic Wind Destruction", (1940)., Engineering News Record, Vol. 125, pp. 672-673.

Farquharson, F. B., (1940), "Dynamic Models of Tacoma Narrows Bridge." Civil Engineering, Volume 10, September, pp. 445-447.

"Tacoma Narrows Bridge Being Studied by Model", (1940), Engineering News

Record, Vol. 125, pp. 139.

"Tacoma Narrows Bridge Wrecked by Wind", (1940), Engineering News Record, Nov. 14, pp. 647.

"Tacoma Narrows Collapse", (1940), Engineering News Record, Vol. 125, pp. 808.

"Model Tests Showed Aerodynamic Instability of Tacoma Narrows Bridge", (1940), Engineering News Record, Vol. 125, pp. 674.

"Resonance Effects of Wind", (1940), Engineering News Record, Vol. 125, pp. 808.

Ammann, Othmar H., Theodore von Karman and Glenn B. Woodruff, (1941), "The Failure of the Tacoma Narrows Bridge, A Report to the Administrator", Federal Works Agency, Washington D.C., March 28.

"Why the Tacoma Narrows Bridge Failed", (1941), Engineering News Record, Vol. 126, pp. 743-747.

Andrew, Charles E., (1943), "Unspinning Tacoma Narrows Bridge Cables", Engineering News Record, January 14, pp. 103-105.

"Tons of Scrap Wire from Tacoma Narrows Bridge", (1943), Engineering News Record, Mar. 11, pp. 87.

Andrew, Charles E., (1943), "Observations of a Bridge Cable Unspinner," Engineering News Record, August 26, pp. 89-91.

Andrew, Charles E., (1945), "Redesign of Tacoma Narrows Bridge", Engineering News Record, Nov. 29, pp. 64-67,69.

Farquharson, F.B., (1946), "Lessons in Bridge Design Taught by Aerodynamics Studies", Civil Engineering, Vol. 16, Aug., pp. 344-345.

Farquharson, F.B., (1950), "Aerodynamic Stability of Suspension Bridges with Special Reference to the Tacoma Narrows Bridge, Investigations Prior to 1941", University of Washington Engineering Experiment Station Bulletin 116, Part I.

Smith, Frederick C., George S. Vincent, (1950), "Aerodynamic Stability of Suspension Bridges with Special Reference to the Tacoma Narrows Bridge, Mathematical Analysis", University of Washington Engineering Experiment Station Bulletin 116, Part II, October.

Vincent, George S., (1951), "Suspension Bridge Vibration Formulas", Engineering News Record, Jan. 11, pp. 32-34.

Farquharson, F.B., (1952), "Aerodynamic Stability of Suspension Bridges with Special Reference to the Tacoma Narrows Bridge, the Investigation of the Models of the Original Tacoma Narrows Bridge under the Action of Wind", University of Washington Engineering Experiment Station Bulletin 116, Part III, June.

Farquharson, F.B., (1954), "Aerodynamic Stability of Suspension Bridges with Special Reference to the Tacoma Narrow Bridge, Model Investigations which

Influenced the Design of the Tacoma Narrows Bridge", University of Washington Engineering Experiment Station Bulletin 116, Part IV, April.

Vincent, George S., (1954), "Aerodynamic Stability of Suspension Bridges with Special Reference to the Tacoma Narrows Bridge, Extended Studies: Logarithmic Decrement Field Damping, Prototype Predictions, Four Other Bridges", University of Washington Engineering Experiment Station Bulletin 116, Part V, 1954.

Goller, R.R., (1965), "The Legacy of Galloping Gertie 25 Years Later", Civil Engineering, ASCE, Vol. 35, Oct., pp. 50-53.

PEACE RIVER BRIDGE - (1957)

The Peace River Bridge on the Alcan Highway in British Columbia failed on October 16, 1957, when the north concrete anchorage block moved forward some 3.7 m (12 ft) on its shale base.

The Peace River Bridge was part of a rush wartime program to complete the Alcan Highway connecting the United States with Alaska. The suspension bridge had a main span of 283 m (930 ft) and the sidespans between the towers and cable bents were 142 m (465 ft) each. Simple truss spans connected the cable bents to the anchorages. The roadway was 7.3 m (24 ft) wide and the center-to-center spacing of the cables was 9 m (30 ft). The cables, made of twenty four 5 cm (1 7/8 in.) strands, were arranged in rectangular form with dimensions of 15 cm by 10 cm (6 in. by 4 in.). The stiffening trusses were 4 m (13 ft) deep.

The Bridge was designed and constructed by the Bureau of Public Roads, the predecessor of the Public Roads Administration. Because of the rush nature of the job, no piling was used to support the anchorages. The sliding of the anchorage on the shale base caused slacking off of the main cables, tipped over the cable bent, dropped the side span suddenly, ripping loose from its 6 cm (2.5 in.) hangers. The first indication that the anchorage was moving came about 12 hours before the collapse when the water supply line crossing the bridge of the new scrubbing plant of the Pacific Petroleum Company was cut. The bridge was immediately closed to traffic. A large crowd gathered to witness the collapse, which was thoroughly photographed. The Canadian Army Engineers put a small ferry 16 km (10 miles) downstream to provide essential transportation for Yukon and Alaska.

Lessons Learned:

In order for a suspension bridge to support the applied loads in the intended manner, it is essential that the anchorages be securely fixed to the ground. Any horizontal motion of an anchorage will cause slackening of the cables with the possibility of collapse of the structure. The Peace River anchorages were supported on footings which did not stay fixed at the intended location.

References:

Birdstall, Blair, (1944), "Construction Record Set on Alcan Suspension Bridge", Engineering News Record, January 13, pp. 26.

"Anchorage Slip Wrecks Suspension Bridge", (1957), Engineering News Record, October 24, pp. 26.

"New Bridge for Peace River on Alcan Highway Planned", (1958), Engineering News Record, January 23, pp. 17.

"Peace River Bridge Failure Was Avoidable", (1958), Engineering News Record, February 13, pp. 10, 16.

"Canada Works on Its Two Fallen Bridges", (1959), Engineering News Record, March 19, pp. 49.

THE SECOND NARROWS BRIDGE - (1958)

On June 17, 1958, the falsework of the partly completed Second Narrows Bridge in Vancouver, British Columbia buckled and plunged two spans of the bridge into Burrard Bay. Fifteen men died in the crash and twenty were injured. The six-lane cantilever truss bridge was to be an important link between the cities of Vancouver and North Vancouver. The main cantilever structure was 620 m (2034 ft) long consisting of a 335 m (1,100 ft) cantilever span, and two 142 m (467 ft) anchor spans. In addition the bridge had four 87 m (285 ft) steel truss and nine 37 m (120 ft) prestressed concrete approach spans. The construction of the bridge began in February 1956 and was to be completed by the end of 1958 at a cost of $16 million.

The two sections that fell were the partly erected north anchor span and a completed simple truss span adjacent to it. The workers were moving additional steel to the overhanging end when the collapse occurred. The bent supporting most of the 20 MN (2,000 ton) anchor span buckled, dropping one end of the span into the water. The impact moved the top of the permanent concrete pier by a few feet, plunging the adjacent simple span into the water.

Lessons Learned:

To determine the reason for the failure of the temporary supports, an investigation was carried out under the British Columbia's Supreme Court Chief Justice, Sherwood Lett. The investigation revealed that the bent supporting the cantilever bridge section was not properly designed. The grillage was designed by comparatively inexperienced engineers without effectively checking the calculations. The bridge was completed in July 1960.

References:

"What Happened at Vancouver? Probers Seek Cause of Bridge Collapse", (1958), Engineering News Record, June 26, pp. 21-22.

"What Happened in Vancouver?", (1958), Engineering News Record, July 3, pp. 100.

"Bridge Crash Witnesses Testify," (1958), Engineering News Record, July 31, pp. 24.

"Faulty Grillage Felled Narrows Bridge," (1958), Engineering News Record, October 9, pp. 24.

"Well Done, British Columbia," (1958), Engineering News Record, Oct. 23, pp. 100.

"Report Blames Contractor for Vancouver Failure," (1958), Engineering News Record, December 11, pp. 32.

"Vancouver Bridge Failure Probed Properly", (1958), Engineering News Record, December 18, pp. 100.

"Vancouver Spans Near Completion", (1960), Engineering News Record, June 30, pp. 24.

KING STREET BRIDGE - (1962)

The King Street Bridge was an all-welded steel girder structure consisting of three main sections, a high level section and two lower level spans which flanked both sides of the high level portion. The spans served to carry roadways over the Yarra River and were completed on April 12, 1961. On the morning of July 10, 1962, brittle fracture failure occurred at points 4.9 m (16 ft) from the ends of one of the 30 m (100 ft) long approach spans under a load of 470 kN (47 tons), which was within the permissible design limits for the bridge, at a temperature of -1° Centigrade (30° Fahrenheit). Three of the four girders fractured at points 4.9 m (16 ft) from both the southern and northern ends whereas the fourth one failed only at one position, namely 4.9 m (16 ft) from the southern end. The failure of the four girders was attributed to a combination of three factors: inappropriate steel for welding, unsatisfactory design details and low ambient temperatures.

The steel used, British Standard 968.1961, is similar to ASTM A 440 and was commonly used in riveted and bolted construction. Welding of such high carbon steel often results in weaknesses being generated in the heat affected zones and the triggering of lamellar tearing failures. Lack of preheating in the short transverse welds at the ends of the cover plates which terminated at the position of fracture is thought to have contributed to crack initiation.

The thickening of the flanges at the points of maximum tensile stress by the addition of cover plates was not a favorable design feature. The temperature on the day of collapse was below that at which the transition from ductile to brittle steel characteristics occur. Brittle behavior favors crack initiation and propagation by increasing the stress intensity factor at any surface or interior flaws. Such conditions were found to contribute to the failure of the Kings Street bridge. The bridge was repaired invoking a scheme of prestressing the girders with steel cables.

Lessons Learned:

Inappropriate steel selection, undesirable design details and unusually low temperatures were the main contributory factors leading to the failure of the Kings Street Bridge. While the low temperature could not have be avoided, the other aspects were within the control of those involved in the bridge's inception.

References:

"Steel Blamed in Bridge Failure", (1962), Engineering News Record, 169, 12, Sept. 12, pp. 139.

"Brittle Fracture of an Alloy Steel Bridge", (1962), Engineering, 194, 5031, Sept. 21, pp. 375.

POINT PLEASANT BRIDGE - SILVER BRIDGE - (1967)

The Point Pleasant I-bar suspension bridge between Point Pleasant, West Virginia and Kanagua, Ohio which was built in 1928, failed at 5:00 p. m. on December 15, 1967. Forty-six people died in the accident and thirty-seven vehicles on the bridge fell with the bridge.

The center span was 213 m (700 ft) long and the side spans were 116 m (380 ft) each. The bridge was unique in that the stiffening trusses of both the center span and the two side spans were framed into the eyebar chain to make up part of the stiffening truss.

Investigation of the failure indicated that collapse of the Point Pleasant Bridge was caused by a defective I-bar at joint 13 of the north chain, approximately 15 m (50 ft) west of the Ohio Tower. The bar which connected Joint 11 to Joint 13 developed a cleavage fracture in the lower portion of its head. Once the continuity of the suspension system was destroyed the bridge collapsed suddenly.

Lessons Learned:

An I-bar suspension bridge is not a fail-safe structure. Failure of one I-bar is sufficient to cause collapse. If a bridge of this type is to be constructed, close and frequent inspection of the structure is necessary.

The tragedy of the failure of the Pleasant Point Bridge led to the national policy for bridge inspections. In 1968 the United States Congress enacted the National Bridge Inspection Standards (NBIS).

References:

Steinman, D.B., (1924), "Design of Florianopolis Suspension Bridge", Engineering News Record, November 13, pp. 780-782.

Ballard, W.T., (1929), "An Eyebar Suspension Span for the Ohio River", Engineering News Record, June 20, pp. 997-1001.

"Disaster", (1967), Time Magazine, December 22, pp. 20.

"Collapse May Never be Solved", (1967), Engineering News Record, December 21, pp. 69-71.

"Point Pleasant Bridge Failure Triggers Rash of Studies", (1968), Engineering News Record, January 4, pp. 18.

"Possible Key to Failure Found", (1968), Engineering News Record, January 11, pp. 27-28.

"Collapsed Silver Bridge is Reassembled", (1968), Engineering News Record, April 25, pp. 28-30.

"Bridge Failure Probe Shuts Twin", (1969), Engineering News Record, January 9, pp. 17.

"Collapse of US 35 Highway Bridge, Point Pleasant, West Virginia, December

15,1967 Highway Accident Report", (1968), National Transportation Safety Board, Washington, D. C., October 4.

Shermer, Carl, (1968), "Eye-Bar Bridges and the Silver Bridge Disaster", Engineers Joint Council, Vol. IX, No. 1, January-February, pp. 20-31.

"Collapse of US 35 Highway Bridge, Point Pleasant, West Virginia, December 15, 1967 Highway Accident Report", (1970), National Transportation Safety Board, Washington, D. C. December 16.

Dicker, Daniel, (1971), "Point Pleasant Bridge Collapse Mechanism Analyzed", Civil Engineering, ASCE, July, pp. 61-66.

Scheffey, Charles F., (1971), "Point Pleasant Bridge Collapse, Conclusions of the Federal Study", Civil Engineering, ASCE, July, pp. 41-45.

Bennett, J. A. and Mindlin, Harold, (1973), "Metallurgical Aspects of the Failure of Point Pleasant Bridge", Journal of Testing Evaluation, ASTM, Vol. 1, No. 2, May, pp. 152-161.

ANTELOPE VALLEY FREEWAY INTERCHANGE - (1971 & 1994)

A few structural failures can be considered milestone in that they have a far reaching impact on design codes and construction techniques. One such failure was the collapse of the Freeway 5/14 south connector overcrossing during the 1971 Sylmar earthquake. The overpass was in the final stage of construction, of pre-stressed concrete box girder design, 411 m (1349 ft) long over nine spans, having a section 10.3 m (34 ft) wide and 2.1 m (7 ft) deep. The longest column of the overpass which was 42.7 m (140 ft) high, had an octagonal section 1.8 m by 3 m (6 ft by 10 ft) with no enlargement where the column intersected with the bottom of the beam section. The foundation for this column consisted of a 6 m (20 ft) deep, 2.4 m (8 ft) diameter cast in place drilled concrete shaft founded onto bedrock. Reinforcement for the column consisted of 52 number 18 bars longitudinally, tied by number 4 bars at 30 cm (12 in.) on centers. This column supported the center of a 117 m (384 ft) long section of the overpass which was connected to the rest of the bridge by way of two shear key type hinges on both ends of the box girder section. The shear keys were 17.8 cm (7 in.) deep vertically and 35 cm (14 in.) long. The sections were also tied together by three 3.8 cm (1.5 in.) steel diameter bolts that were added to equalize the longitudinal deflections in the superstructure arising from creep and temperature effects. This section was different from the rest of the bridge because of the fact that it was supported by one column instead of two.

On February 9, 1971 at 6:01 a.m. an earthquake assessed at Richter magnitude 6.6 occurred in the mountains behind Sylmar. The interchange suffered horizontal accelerations that were estimated as high as 0.6g. The 10 to 15 seconds of strong motion caused the superstructure of the 117 m (384 ft) section of the overpass to jump out of the shear key seats and induced the column and bridge deck to act as an inverted pendulum. The capacity of the column was found inadequate and it failed in bending at the base.

It was generally agreed that the overpass was of superior construction and did not fail as a result of any defects in workmanship or construction techniques. Pre-stressing elements survived the earthquake loading well and were intact in the debris.

As a result of this experience significant changes in bridge design criteria were made including very large increases in the seat sizes to allow for much greater longitudinal and lateral horizontal movements, the location of hinges so that there are a least two columns between adjacent hinges along the bridge, the incorporation of spiral reinforcement to confine the steel within the columns, the elimination of lap slices at the base of the columns, the reduction of skew in overpass structures, the increase in the amount of reinforcement at the column stroke deck connection to provide greater resistance to punching shear and the elimination of the use of rocker type bearings.

On January 17, 1992 the Richter Magnitude 6.4 Northridge earthquake again caused failure of portions of this interchange . On this second occasion some of the most severe damage occurred to sections that had been repaired following the 1971 earthquake and in other instances spans that had been under construction in 1971, failed. The fact that some spans were supported on columns of greatly dissimilar heights was thought to have contributed to the failures. The shorter columns, being much stiffer, were considered to have attracted disproportionately large shear forces resulting in their being overloaded with an inevitable subsequent domino effect. Also inadequate scat lengths at the ends of several spans contributed to collapse. Apparently the interchange had been scheduled for a seismic upgrade but the 1992 earthquake occurred before this had been started.

Lessons Learned:

The failure of the I 5/14 interchange in 1971 represented a turning point in seismic design of freeway bridges and prompted a radical change in the seismic design provisions for such structures. However these changes were not applied to the I 5/14 interchange itself. The failure in 1992 reemphasized the dangers of procrastination in undertaking seismic retrofitting when once the need for such action has been established.

References:

"Engineering Features of the San Fernando Earthquake, February 9, 1971", (1972), Earthquake Engineering Research Laboratory Report 71-02, June, California Institute of Technology.

"Northridge Earthquake", (1994), Civil Engineering, ASCE, Vol. 64, No. 3, March, pp. 40-47.

"The Northridge, California Earthquake of January 17, 1994: Performance of Highway Bridges", (1994), Technical Report NCEER-94-0008, State University of New York at Buffalo, March.

Antelope Valley - Interstate 5 Freeway Interchange Failure - 1994

MIANUS RIVER BRIDGE - (1983)

The Mianus River bridge forms part of the Connecticut turnpike which is 206 km (129 miles) long and crosses the state from Rhode Island to New York as part of Interstate 95. It was opened in 1958. The bridge is a six lane structure crossing the river at a 53 degree angle and is comprised of cantilever sections, suspension spans and pinhanger connections. The deck is 19 cm (7.5 in.) thick concrete with a 5 cm (2 in.) thick asphalt top carried on steel plate girder construction, the longest span of which is 809 m (2, 656 ft). The cantilever sections are projected from adjacent skewed concrete piers. The three lane suspension spans, 10.6 m by 30.5 m (35 ft by 100 ft), are held in place by two cantilever sections. One such section weighs five hundred tons and includes two nine foot deep plate girders and four stringers connected by a network of cross beams. Each span is connected at the corners by two pins and a 2 m (6.5 ft) long hanger. The pins are 17.8 cm (7 in.) long, 17.8 cm (7 in.) in diameter and each had a 2.5 cm (1 in.) diameter hole along its axis to receive an 28.5 cm (11.25 in.) long steel bolt with washers and a nut welded on each end. A pair of such pins attaches two hangers to the cantilever girder and the suspension girder. Hence, one span has a total of eight pins and eight hangers.

On June 28, 1983, in the early hours in the morning, a 30 m (100 ft) long section of the Mianus River bridge fell taking down with it two tractor trailers and two passenger vehicles. The fallen section was a three lane east bound suspended span near to the Long Island Sound entrance. The adjacent west bound section remained in place. In the investigation which followed the collapse, several of the steel pins were recovered. A portion of the eighth was also found. Metallurgical tests were conducted on the fragmented pin.

An inquiry concluded that a lack of maintenance and inadequate inspection were the main causes of the collapse. It is believed that corrosion caused the a pin to be pushed off the hanger which triggered the collapse.

As a result of the Mianus River bridge incident, other bridges with similar design were inspected thoroughly. It was found that four other bridges on the turnpike were suffering similar deterioration thereby confirming the failure to maintain the hanging mechanism on these bridges had developed into a serious problem in other cases than the one collapsed section of the Mianus bridge.

Lessons Learned:

It was clear from this event that inadequate maintenance of major bridges can lead to catastrophic collapse. The failure of the Mianus River bridge was an example of the neglect of the infrastructure which has become a major concern to civil engineers in recent years and emphasizes the need for not only ensuring adequate standards of design and construction but also of maintenance and of service throughout the life of all structures. Many repairs and retrofits were initiated to provide increased redundancy in other bridges.

References:

"Mianus River Bridge to Open," (1958), New York Times, July 18, pp. 23.

"Failed Pin Assembly Dropped Span," (1983), Engineering News Record, July 7, pp. 10-12.

"Temporary Span Rushed in.," (1983), Engineering News Record, July 7, pp. 11-12.

"Girder Flaws Eyed in Span Collapse," (1983), Engineering News Record, July 14, pp.12.

"Hearings on Collapsed Span Focus on Rust, Skewed Plan," (1983), Engineering News Record, Sept. 29, pp. 35-36.

"Engineers Say Faulty Design was Factor in Mianus Bridge Collapse," (1983), New York Times, July 18, pp. 23.

"New Mianus River Bridge Report Disputes Earlier Study," (1985), Civil Engineering, ASCE, April, pp. 10.

"Designers Cleared in Bridge Collapse," (1986), Civil Engineering, ASCE, April pp. 10.

Mianus River Bridge Collapse

SAN FRANCISCO-OAKLAND BAY BRIDGE - (1989)

The two deck San Francisco/Oakland Bay bridge was opened in 1936 and comprises two distinct structures. The West Bay Crossing from San Francisco to Yerba Buena Island is a twin suspension structure, whereas the East Bay Yerba Buena to Oakland portion is composed of a series of simple span trusses and a long cantilever truss. The seismic design was based on a coefficient of 0.1g in accordance with the standards at the time the design was completed.

On October 17, 1989 the 7.1 Richter magnitude Loma Prieta earthquake was centered about 100 km (60 miles) south of the San Francisco/Oakland bridge but caused a 15 m (50 ft) span of the upper deck to collapse onto the lower one, resulting in the death of one motorist who had the misfortune to drive into the gap.

The 4 km (2.4 mile) East Bay Crossing uses 10 bridge piers as anchor points, the longitudinal forces being transferred out of the structure at these positions. These bridge piers vary in construction and the one where the span collapsed is a four column diagonally braced steel tower. It services a tributary distance of 775 m (2,544 ft) to the west and 193 m (632 ft) to the east. Since the tributary distance to west of the pier is much the larger, the result of longitudinal forces from the west will normally be of a greater magnitude then those from the east. To connect the two truss system together an expansion joint system consisting of I beam stringers resting on 15 cm (6 in.) wide stiffened seat supports at the west end and bolted stringer flanges to stiffened seat connections at the east end were used. The stiffened seats provided 13 cm (5 in.) of bearing support for the stringers in the precollapsed configuration. The upper deck floor system used four stringers which supported transverse joists which carried the concrete deck. The lower deck floor system used eleven stringers which directly supported the concrete deck. It has been estimated that during the earthquake the maximum longitudinal acceleration suffered by the eastern portion was 0.22g. The longitudinal inertia force arising from this acceleration caused the twenty-four 2.5 cm (1 in.) diameter bolts used to anchor the fixed shoes to the tower columns to shear off. The truss was then free to move with the motions of the earthquake and as the deck moved eastward the stringers were pulled away from the west truss, exceeding the bearing distance of the stiffened seat and causing the stringers to fall at the west end. The actual easterly displacement of the shoes was measured to be at least 17.8 cm (7 in.), judging from the marks on the base plates to which the shoes had previously been attached.

The bridge was restored to use within a month of the earthquake but controversy raged for many years with respect to the optimum method of retrofitting in order to provide greater seismic resistance than that possessed by the original structure.

Lessons Learned:

The loss of a major lifeline transportation link in the San Francisco Bay area demonstrated the vulnerability of a integrated economic area to the loss of a major artery. It prompted the application of state of art methods of earthquake engineering analysis to evaluate what seismic forces and movements long bridges are likely to experience in major earthquakes and focused attention on the need to retrofit many of the major bridges in seismic prone areas.

References:

Astaneh, A., (1989), "Preliminary Report on the Seismological and Engineering Aspects of the October 17, 1989 Earthquake", Report No. UCB/EERC-89/14, Earthquake Engineering Research Center, University of California at Berkeley, Oct., pp. 34-37.

"Loma Prieta Reconnaissance Report", (1990), Supplement to Earthquake Spectra, 6, Earthquake Engineering Research Institute, pp. 162-169.

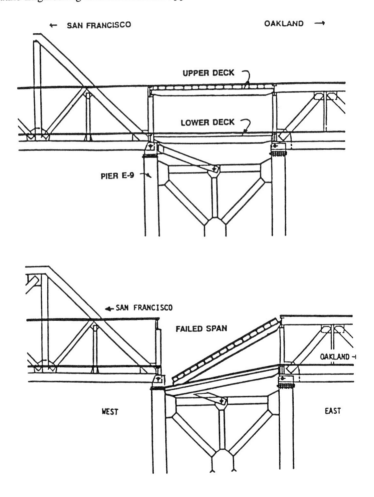

Detail of Collapse of Span of San Francisco - Oakland Bay Bridge

CYPRESS VIADUCT - (1989)

The Cypress Viaduct was California's first continuous double-deck freeway structure. Construction commenced in 1955 and it was opened to traffic in June, 1957. Situated just west of downtown Oakland, the structure extended approximately 2.4 km (1.5 miles) in a north-south direction. Both the upper, southbound, and lower, northbound, traffic lanes were supported above ground level by a series of reinforced concrete bents. The upper frame incorporated shear keyed, essentially pinned, joints positioned so that the upper part of each bent was statically determinate.

Some recognition of the potential hazard posed by a structure which had been designed to much lower seismic load demands than would be appropriate for a structure conceived more recently had resulted in some retrofitting of the viaduct being undertaken. This work comprised essentially tying the spans together longitudinally. No strengthening of the joint areas around the shear keys was done.

The majority of the total fatalities which occurred as a result of the 1989 Loma Prieta earthquake resulted from the collapse of the upper deck onto the lower elevated one, trapping and crushing vehicles in the northbound lanes. The fact that other nearby bridges and buildings survived the seismic shaking focused attention on the configuration of the Cypress Viaduct, the details of the structure and the site on which it was founded. An extensive investigation was mounted by many groups of researchers including those from the University of California, Berkeley, as a result of which the characteristics of the structure and its behavior in the October 17, 1989 earthquake were clarified.

The majority of the collapsed bents followed a common pattern of failure involving slipping along a plane of weakness which existed in the stub region of the lower column to beam joint, just below the shear key at the bottom of the upper bent column. This plane of weakness in the joint region was created by closely spaced lower girder negative moment steel reinforcement which was bent down into the column. Insufficient transverse reinforcement was provided to prevent the wedge of concrete outside the plan of the bent down girder reinforcement from sliding on the sloping failure surface under the combined effects of the lateral seismic loads and the weight of the upper deck.

Lessons Learned:

The reinforcement in the columns and girders was poorly detailed, even allowing for the lack of understanding, at the time the viaduct was designed, of the inelastic response demands on reinforced concrete structures in seismic zones. Confirmation of the necessity to design for realistic earthquake generated forces and displacements was one of the lessons learned from the experience of the Cypress viaduct collapse. Another was the penalty of delaying seismic strengthening when once it is established that a structure is not up to current design standards.

References:

"Collapse of the Cypress Street Viaduct as a Result of the Loma Prieta Earthquake", (1989), Report No. UCB/EERC-89/16, Earthquake Engineering Research Center, University of California at Berkeley.

Housner, G.W., (1989), "Competing Against Time", Report to Governor George Deukmejian on 1989 Loma Prieta Earthquake, 264 pp.

"Loma Prieta Reconnaissance Report", (1990), Supplement to Earthquake Spectra, 6, Earthquake Engineering Research Institute, pp. 151-159.

Cypress Viaduct Collapse

BUILDING FAILURES

AMC WAREHOUSE - (1955)

The AMC Warehouse at Wilkins Air Force Depot, Shelby, Ohio consists of 122 m (400 ft) long six-span rigid concrete frames spaced at 10.6 m (35 ft) on-center along the 61 m (200 ft) length of the building. The 10 cm (4 in.) cast-in-place gypsum roof slab is supported by prestressed concrete block purlins spanning between the rigid frames. An expansion joint runs the length of the building near the center of the 122 m (400 ft) wide frames.

Due to distress noted during construction of similar warehouses at other Air Force facilities, the design of the Wilkins warehouse was modified during construction in March of 1954. The revision included the addition of continuous full length top bars and nominal shear stirrups throughout the length of the span. Approximately 18 months after casting, diagonal cracks were first noticed in the concrete frames. In August of 1955, three adjacent concrete frames collapsed under dead load of the frames and roof only. All frames failed about 45 cm (18 in.) beyond the cut-off location of the negative reinforcement in the first interior span. The collapsed area was approximately 370 sq m (4000 sq ft).

Lessons Learned:

The collapse was attributed to insufficient web reinforcement and insufficient extension of both positive and negative reinforcing steel. The reinforcement was not able to resist the combined flexural, shear and longitudinal stresses. The longitudinal stresses (shrinkage and temperature) were believed to be higher than expected due to ineffective expansion joints. A series of load tests were performed on one-third scale model specimens of the concrete frames at the Portland Cement Association. The laboratory study was able to replicate the type of failure observed in the structure to verify the cause of the collapse. All the frames were reinforced with vertical steel straps placed around the girders of the frames to increase their shear capacity. The repair scheme was also tested at the Portland Cement Association laboratory.

Since the failure investigation found that the design of the structure conformed to current codes, the ACI Building Code was revised to limit the allowable shear stress in members without web reinforcement to the lesser of 0.03 f'c or 620 kPa (90 psi). In addition, the code revision required that web reinforcement for continuous or restrained beams (except T-beams with an integral slab) be extended from the support to a point either 1/16 of the clear span or the depth of the member, whichever is greater, past the extreme point of inflection. The web reinforcement was also required to carry at least two-thirds of the total shear in this region and at least two-thirds of the total shear at any section which contained negative moment reinforcement.

References:

Anderson, B.G., (1957), "Rigid Frame Failures," Journal of the American Concrete Institute, Vol. 28, No. 7, pp. 625-636.

Elstner R.C. and Hognestad, E., (1957), "Laboratory Investigation of Rigid Frame Failure," Journal of the American Concrete Institute, Vol. 28, No. 7, pp. 637-668.

Lunoe R.R. and Willis, G.A., (1957), "Application of Steel Strap Reinforcement to Girders of Rigid Frames, Special AMC Warehouses," Journal of the American Concrete Institute, Vol. 28, No. 7, pp. 669-678.

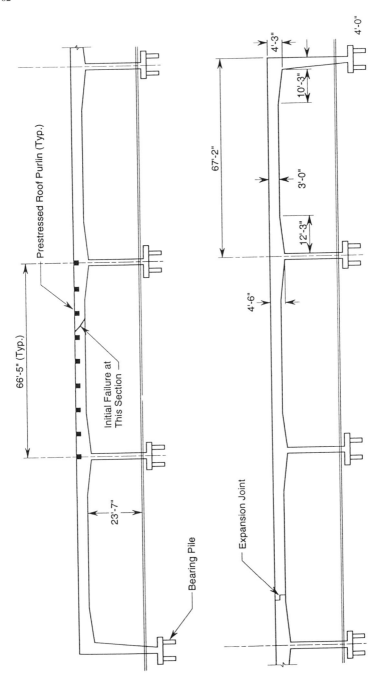

Detail of Typical Frame at AMC Warehouse

RONAN POINT TOWER - (1968)

The need to provide replacement housing for that destroyed in the 1939-45 war prompted the development in Europe of innovative prefabricated construction techniques in the decades following. One such scheme involved the erection of high rise apartment buildings using factory made concrete components. The structural system comprised load bearing walls, with each level of apartments stacked directly on the one below. Floor on wall and wall on floor joints were grouted bearing surfaces.

On May 16, 1968 an undetected gas leak resulted in an explosion in the kitchen of a unit on the eighteenth floor when the occupant attempted to light the stove. The corner walls of this unit blew out, causing the wall above to collapse. These in turn impacted on the floors below and destroyed the whole corner of the building. Fourteen people were injured, three fatally.

Analysis of the event reveled the disadvantage of no alternative load paths being present when one part of an external wall at one level was removed. Demolition of the building also confirmed that deficiencies existed in the quality of the grouted joints between the prefabricated components.

The outcome of the Ronan Point Tower episode was that doubt was cast on the safety of other apartment towers using similar structural systems. Many were demolished well in advance of their expected life expectancy. Progressive collapse, in which local failure is followed by a chain reaction producing widespread collapse, certainly was not unknown prior to the Ronan Point event. Structures are particularly susceptible to this domino effect in the course of the construction process. What was unusual in the case of Ronan Point was that a relatively minor gas explosion triggered the collapse of a significant portion of a completed building.

Lessons Learned:

The experience of Ronan Point reemphasized the need to be aware of the possibility of progressive collapse of constructed facilities, the desirability of providing redundancy - or fail safe possibilities - in structural systems and the necessity of ensuring quality control in the construction process.

References:

"Report of the Inquiry into the Collapse of the Flats at Ronan Point, Canning Town", (1968), HMSO, London.

Allen, D.E. and Schriever, W.R., (1972), " Progressive Collapse, Abnormal Loads and Building Codes", ASCE, Structural Failure: Modes, Causes, Responsibilities.

Kaminetzky, D., (1991), "Design and Construction Failure", McGraw Hill

Levy, M. and Salvadori, M., (1992), "Why Buildings Fall Down", Norton and Co., New York, Chapter 5.

Petroski, H., (1994), "Design Paradigms", Cambridge University Press.

Ronan Point Tower

THE SKYLINE PLAZA APARTMENT BUILDING - (1973)

The Skyline Plaza complex in Fairfax County, Virginia, included eight apartment buildings, six office buildings, a hotel and shops. Two of the apartment buildings had been completed, with another pair under construction, when one of the buildings under construction collapsed. Fourteen construction workers were killed, and thirty-four more were injured, on that fateful day of March 2, 1973.

The apartment building was to consist of twenty-six stories, plus a penthouse and a four story basement. A parking structure was attached. The apartments consisted of reinforced concrete flat plate construction with 20 cm (8 in.) concrete floor slabs and a typical story height of 2.7 m (9 ft).

On Friday, March 2, 1973, at approximately 2:30 p.m., a portion of one of the apartment buildings under construction collapsed. The collapse started with a section of the twenty-third floor slab, under the twenty-fourth floor slab being cast, and proceeded vertically the full height plus basement levels, as well as destroying the adjacent parking facility.

Immediately after the incident, an inspection team from the Occupational Safety and Health Administration (OSHA) began an investigation of the site. A detailed investigation involving OSHA and the Center for Building Technology of the National Bureau of Standards followed.

Lessons Learned:

The investigation revealed serious violations of specified construction requirements (non-compliance with OSHA construction standards), including:
• Violation of requirement to fully-shore the two floors beneath the floor being cast,
• Failure to allow proper curing time before removal of shores (premature removal of twenty- second story forms).
• Failure to prepare or test field cured concrete specimens.
• Shoring damaged and/or out of plumb.
• Failure of inspection to note or correct violations.
• Improper "climbing" crane installation.

References:

Leyendecker. E. V., and Fattal, S. G., (1977), "Investigation of the Skyline Plaza Collapse in Fairfax County, Virginia," National Bureau of Standards, Building Science Series Number 94, February.

Dixon, D. E., and Smith, J. R. (1980), "Skyline Plaza North (Building A-4) A Case Study," ASTM Special Technical Publication 702, Full-Scale Load Test of Structures, April 2, 1979 Symposium, Philadelphia, Pa. pp. 182-199.

Carino, N. J., Woodward, K. A., Leyendecker, E. V., and Fattal, S. G., (1983), "Review of the Skyline Plaza Collapse", Concrete International Design and Construction, Vol. 5, Number 7, July, pp. 35-42.

Skyline Plaza Apartment Construction Failure

HARTFORD CIVIC CENTER COLISEUM - (1978)

The $75 million civic center in Hartford, Connecticut, which housed a coliseum, retail shops, and convention space, was completed in 1973. The coliseum roof collapsed before dawn on January 18, 1978, after a snowstorm, and just hours after the last fans left a well-attended basketball game.

The structure consisted of a 10,000 sq m (108,000 sq ft) space truss roof, which was near record size for its time. The 7.6 m (25 ft) high structure was 110 m by 91 m (360 ft by 300 ft), with clear spans of 64 to 82 m (210 to 270 ft). It consisted of Warren trusses with triangular bracing between top and bottom chords. Struts 1 m (3 ft) long attached to the space truss were used to support the 7.5 cm (3 in.) wood fiber composition roofing. The struts were not crossbraced. The space truss had been assembled on the ground and jacked into place. The original design and analysis was done via computer.

During the night prior to the collapse there was a heavy snowstorm, which changed to rain. The roof began to sag, top chords of the truss buckled, redistributing the load to other members, the roof folded, and it ultimately collapsed, dropping 25 m (83 ft) to the 12,000 seat arena below.

Lessons Learned:

Investigation of the failure indicated that although the design considered midspan bracing of the top chord members, construction details showed those chords to be unbraced over their 9 m (30 ft) length. The use of the struts without cross-bracing to carry the roofing minimized the diaphragm action of the roof. The roof dead loads were also seriously underestimated. Based upon the above, the roof was extremely susceptible to a buckling failure, a mode not considered in the computer design.

In addition, the investigation revealed that excessive deflection occurred during the construction phase (significantly exceeding the design deflection), which should have been noted during inspection, alerting the design professionals to the inadequacies of the design and the potential for catastrophic failure.

This failure has been appropriately referred to as a "computer-aided catastrophe". It illustrates the false sense of security offered by the multitude of data generated by computer analysis. It serves as a lesson that computer software is a tool, and not a substitute, for sound engineering experience and judgment. Relying on computer analysis only for the analysis of increasingly complex structures, where the reasonableness and thoroughness of the analysis cannot be ascertained, invites failure.

References:

"Space Frame Roofs Collapse Following Heavy Snowfalls", (1978), Engineering News-Record, January 26, pp. 8.

"Probe Closes in on Why Space Frame Roof Failed", (1978), Engineering News-Record, March 16, pp. 13.

"Design Fault Suspected in Hartford Failure", (1978), Engineering News-Record, March 30, pp. 3.

"Design Flaws Collapsed Steel Frame Roof", (1978), Engineering News-Record, April 6, pp. 9.

"Someone Should Have Sounded Alarm", (1978), Engineering News-Record, April 6, pp. 88.

"Space Truss, Not Space Frame", (1978), Engineering News-Record, June 22, pp. 29.

"Collapsed Space Truss Roof had a Combination of Flaws", (1978), Engineering News-Record, June 22, pp. 36.

"Collapsed Roof Design Defended", (1978), Engineering News-Record. June 29, pp. 13.

"$12.3-Million Settlement in Roof Collapse", (1978), Engineering News-Record, July 6, pp. 3.

"Design, Inspection Blamed in Roof Collapse", (1978), Engineering News-Record, July 27, pp. 15.

"Hartford Coliseum Suit", (1978), Engineering News-Record, November 30, pp. 10.

"On Collapse Report Authorship", (1978), Engineering News-Record, September 7, pp. 3.

"Hartford: Designer's Duties", (1978), Engineering News-Record. September 21, pp. 21.

Hartford Civic Center - Pre Failure

Hartford Civic Center - Post Failure

IMPERIAL COUNTIES SERVICES BUILDING - (1979)

The six story reinforced concrete Imperial Counties Services was only eight years old when it was crippled by the 1979 Imperial Valley Richter magnitude 6.6 earthquake.

The structure comprised a five story box-like, relatively rigid, upper portion of approximately 930 sq m (10,000 sq ft) footprint, that was supported on columns and irregularly placed shear walls at the first floor level.

Severe damage was experienced by the columns at the east end of the building with bursting of the rebar ties and concrete crushing resulting in significant shortening. The building was judged unrepairable and was subsequently demolished.

The exception circumstance of this building was the presence of a thirteen transducer array of strong motion recorders which provided, for the first time ever, a set of data recording the behavior of a building which received major structural damage. This fact, coupled with the availability of measurements made of the significant dynamic characteristics prior to the earthquake, allowed detailed analyses to be undertaken of the sequence of the failure.

Lessons Learned:

The fact that architectural configuration has an important influence on the ability of a building to resist earthquakes was confirmed by the behavior of the Imperial County Services Building in the 1979 earthquake.

References:

Reconnaissance Report, Imperial County, California, Earthquake of October 15, 1979, (1980), Earthquake Engineering Research Institute.

Shepherd, R., and Plunkett, A.W., (1983), "Analysis of the Imperial County Services Building", Journal of Structural Engineering Division, ASCE, Vol. 109, No. ST7, pp. 1711-1726.

Imperial Counties Services Building

KEMPER MEMORIAL ARENA ROOF - (1979)

A significant portion of the roof of the 17,600-seat Crosby Kemper Memorial Arena in Kansas City, Missouri collapsed on June 4, 1979. The collapse occurred at approximately 7:10 P.M. during a severe wind and rain storm. Of the 12,000 sq m (130,000 sq ft) roof, approximately 4,000 sq m (43,000 sq ft) had to be replaced. There were no deaths or injuries; only two maintenance personnel were present at the time of the failure.

The arena is a reinforced concrete seating and service structure, enclosed by metal wall panels and roof hung from structural steel pipe space frames. Three individual triangulated space frames span the short dimension and support the secondary roof structure (steel plane trusses) by pipe hangers at 42 different panel points. The space frames are made of large steel pipe members bolted together with A-490 bolts loaded in shear. The hangers were fastened to the space frames by a welded gusset plate with a pin connection, and to the top chords of the secondary roof trusses by a steel plate welded to the hangers, then bolted with four A-490 bolts, vertically loaded in tension.

Wind loading during the five years of occupancy of this building caused many cycles of loading and unloading of the bolts in the tension connection. These dynamic effects were determined to be the most probable cause of collapse. A contributing factor was the way in which the connection was constructed, in that considerable movement through prying action was permitted. High-strength steel bolts, which are relatively brittle, are not recommended for conditions other than static loading. This failure illustrated the reasons for that recommendation and the desirability of ductile connections. Also cited was the lack of redundancy in the basic design - a common problem in modern long span structures. In this case, fatigue failure in a single bolt was able to bring about a progressive collapse of the entire center section of the roof and significant damage to the rest of the structure.

Lessons Learned:

Several dramatic structural failures occurred within a couple of years. These included the Hartford Connecticut Civic Center Coliseum roof (1978), Kansas City Kemper Arena (1979), and the Hyatt Regency Hotel walkways (1981). Taken together, these failures had a sobering effect on the construction industry. Attention was focused on the need to provide greater structural integrity and redundancy in the design of structures. In particular, the importance of connection design was emphasized. Detailing and execution of connections, including provisions for their adequate review during the shop drawing phase, has been the subject of significant discussion since these failures occurred. In addition, this failure and others have contributed to the understanding of wind forces on buildings and the evolution of improved wind design standards to consider dynamic effects and pressure concentrations.

References:

Engineering News Record, (1979), "Clean-up Begins as Kemper Designer Sues for Access," McGraw-Hill, New York, NY, June 28, 1979, pp. 13.

Stratta, J.L., (1979), "Report of the Kemper Arena Roof Collapse of June 4, 1979. Kansas City Missouri", James L. Stratta, Consulting Engineer, Menlo Park, CA.

Engineering News Record, (1980), "Kemper Arena Reopens with New Roof Connections," McGraw-Hill, New York, NY, March 13, pp. 18.

Civil Engineering, (1981), "Kemper Arena Roof Collapse and Repair," ASCE, New York, NY, March, pp. 68-69.

Kemper Memorial Arena Roof Collapse
(Civil Engineering, March, 1981; Reprinted by permission of ASCE)

HYATT REGENCY HOTEL PEDESTRIAN WALKWAYS - (1981)

On July 17, 1981, two suspended walkways in the atrium area of the Hyatt Regency Hotel in Kansas City, Missouri collapsed suddenly. The failure caused 114 deaths and 185 injuries. Because of the large number of casualties, this failure had a more significant impact on the design and construction industry than any other failure in recent times.

The hotel had been in service for a full year prior to the collapse. The walkways, suspended by tension rods from the atrium roof structure, were arranged such that the second floor walkway was suspended directly below the fourth floor walkway. A separate walkway at the third floor level was not involved in the collapse.

The most probable technical cause of the failure was easily established. A deficient connection, where the steel suspension rods were connected to box beams that supported the fourth floor walkway, failed. This caused the fourth and second floor walkways - 570 kN (64 tons) of debris - to fall to the floor of the atrium area. Each walkway and the atrium floor were filled with guests attending a tea party and dance.

Tests conducted at the National Bureau of Standards showed clearly that the connections as originally shown on the design drawings were not capable of supporting the gravity load required by the relevant building code. This deficiency was compounded by a change made to the detail during construction, which doubled the load on the connection, making its failure inevitable.

There was very little argument within the industry relative to the technical explanation of this failure. However, discussion about the chain of events that permitted this collapse to occur was extremely significant. Questions regarding deficiencies in the project delivery system were the focus of landmark litigation and administrative hearing decisions. These questions were quite basic: Who designed the original connection? Who is responsible for connection design - the fabricator, or the structural engineer? Was the original connection buildable? Who initiated the change to the connection during the construction phase and why? Who approved it? What is the meaning of "shop drawing review?" This failure led directly to ongoing activities aimed at improving quality assurance and quality control in the design and construction process.

Lessons Learned:

In terms of human suffering, the Hyatt Regency Hotel walkway collapse was one of the most significant tragedies to ever confront the construction industry. The technical cause of the failure was easy to understand. However, there were many lessons for design and construction professionals. The important lessons involved procedural issues. Clearly, there is a need for all parties to understand their responsibilities and to perform their assignments competently. The structural engineer's responsibility for overall structural integrity, including the performance of connections, was firmly established in this case. This failure also lends credibility to the practices of project peer review and constructability review.

References:

NBS, (1982), Investigation of the Kansas City Hyatt Regency Walkways Collapse, NBSIR 82-2465, National Bureau of Standards (National Institute of Standards and Technology), Department of Commerce, Washington, D.C.

Pfang, E. O. and R. Marshall, (1982), "Collapse of the Kansas City Hyatt Regency Walkways," Civil Engineering, ASCE, New York, NY, July, pp. 65-68.

Dallaire, G. and R. Robison, (1983), "Structural Steel Details: Is Responsibility the Problem?" Civil Engineering, ASCE, New York, NY, October, pp. 51-55.

Leonards, G. A., (1983), "FORUM: Collapse of the Hyatt Regency Walkways: Implications," Civil Engineering, ASCE, New York, NY, March, pp. 6.

Robison, R., (1984), "Structural Steel Details: Comments on Divided Responsibility," Civil Engineering, ASCE, New York, NY, March, pp. 58-60.

Rubin, R. A. and L. A. Banick, (1987), "The Hyatt Regency Decision: One View," Journal of Performance of Constructed Facilities, ASCE, New York, August, pp. 161-167.

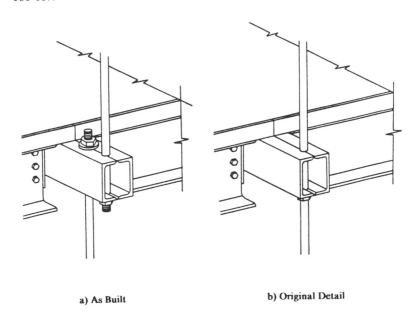

a) As Built   b) Original Detail

Connection Detail - Hyatt Regency Walkway

PINO SUAREZ BUILDING - (1985)

The Pino Suarez was constructed in Mexico City in the early 1970s and consisted of five high rise steel buildings supported on a two level reinforced subway station that acted as a rigid foundation common to all five buildings. One of the functions of the complex was to act as a counterweight against uplift forces caused by the expansive soils which occurred as the consequence of the large excavation for the subway station. In the middle of the complex were three identical 21 story buildings each two bays wide and four bays deep, aligned and parallel, with their narrow sides facing east/west. Directly adjacent to, but north and south of the 21 story buildings were two identical 14 story towers. These two buildings were positioned with their narrow faces directly north and south respectively relative to the longitudinal sides of the 21 story buildings. The structural framing of the 21 story buildings consisted of truss beams connected to hollow box columns with moment resisting connections. In addition a bracing system was employed that consisted of two cross based frames in the transverse direction and one V-braced frame in the exterior longitudinal plane.

As a result of the September 19, 1985 Mexico City earthquake, the complex suffered severe structural failure. One of the three 21 story steel towers collapsed onto an adjacent 14 story tower, destroying it also. The performance of the other two 21 story towers that were severely damaged but did not collapse provided invaluable information with respect to the progression of failure. Computer simulations confirmed the progression of failure beyond reasonable doubt. It was inferred that plastic hinges first developed at the girder ends leading to yielding and plate buckling of the two fourth story columns located on the south side. The accompanying figure illustrates the plate buckling failure in the one of the surviving box columns. Columns failing in this way lost most of their gravity carrying capacity and suffered shortening. This led to buckling of the X-brace framing and the redistribution of forces throughout the structure that eventually caused the failure of the other four story columns and the collapse of the tower.

Lessons Learned:

The Pino Suarez complex failure was caused by a design flaw that was not recognized in the then current seismic design codes. After the failure, Mexico updated its seismic design code recognizing the problem of overloaded columns. The most important lesson is that columns should never be allowed to be overloaded to the point of failure. Additionally, it should be recognized that although member yielding provides a great energy dissipation capability, it should be explicitly considered in design. Sufficient connection strength is necessary and yielding should be confined primarily to beams and bracing not to columns and joints. Finally, the fact that structural redundancy can be advantageous was reemphasized by the failure of the Pino Suarez complex.

References:

Hanson, R.D. and Martin, H.W., (1987), "Performance of Steel Structures in the September 19 & 20, 1985 Mexico Earthquakes," Earthquake Spectra, Vol. 3, No. 2, pp. 329-345.

Osteraas, J. and Krawinkler, M., (1989), "The Mexico Earthquake of September 19, 1985 - Behavior of Steel Buildings", Earthquake Spectra Vol. 5, No. 1, pp. 51-87.

Pino Suarez Building - Buckled Column

L'AMBIANCE PLAZA - (1987)

On April 23, 1987, at approximately 1:30 p.m., the structural frame and slabs of what was to be a sixteen story apartment building in Bridgeport, Connecticut, collapsed during construction, killing 28 construction workers. This was the largest loss of life in a construction accident in the United States since 51 workers were killed in the collapse of a reinforced concrete cooling tower under construction at Willow Island, West Virginia, in 1978.

The L'Ambiance Plaza project comprised two rectangular towers, each with 13 levels of apartments and 3 levels of parking. The towers were being constructed using the lift-slab method. Floor and roof slabs were two-way, unbonded, post-tensioned flat plates. They were all cast at the ground level, post-tensioned and then lifted into place. Steel columns supported the floor slabs, which were lifted in stages by hydraulic jacks, threaded rods, and welded steel shearhead collars placed in the concrete floor slab at each column.

The collapse investigation was complicated by a number of potential contributing factors, and by the unavoidable damage to evidence that occurred during the frantic rescue operation. Several triggering causes were hypothesized, most of these centering on deficiencies in the design and construction of the shearhead collars. One theory proposed that a lifting angle in the shearhead deformed, allowing the nut at the end of the lifting rod to slip out of the assembly. Another theory involved a rolling out of the wedges used to temporarily support slabs during the lifting operation. Deficient welds were also discovered in the shearheads.

Several factors present in the design and construction sequence were identified as contributors to the magnitude of this catastrophe. These included improper placement of some of the prestressing tendons and general overall instability of the frame during construction (shear wall construction was lagging behind the lifting operation.)

A remarkable and somewhat controversial mediated global settlement of claims resulting from this failure brought an end to forensic investigations in December 1988. There may never be complete general agreement on the technical causes of the failure, but one result of the early settlement was that the various theories were discussed thoroughly in the engineering literature, much sooner than has been the case for other collapses of this magnitude.

Lessons Learned:

The tragedy of the L'Ambiance Plaza collapse gave impetus to efforts aimed at ensuring structural integrity during the construction phase of a project. The lift-slab method has an enviable safety record. But this failure underlined the need to carefully evaluate details and sequencing specifically for each project and each site. Much discussion also centered on the convoluted and fragmented project delivery system, in which responsibility for ultimate structural safety was confused by unclear relationships among the engineer of record, the lift-slab contractor, and the designer of the shearheads.

References:

Scribner, C.F. and C.G. Culver, (1988), "Investigation of the Collapse of L'Ambiance Plaza," Journal of Performance of Constructed Facilities, ASCE, Vol. 2, No. 2, pp. 58-79.

Felsen, M.D., (1989), "Mediation that Worked: Role of OSHA in L'Ambiance Plaza Settlement," Journal of Performance of Constructed Facilities, ASCE, Vol. 3, No. 4, pp. 212-217.

Poston, R., G.C. Feldmann and M.G. Suarez, (1991), "Evaluation of L'Ambiance Plaza Post-tensioned Floor Slabs," Journal of Performance of Constructed Facilities, ASCE, Vol. 5, No. 2, pp. 75-91.

Heger, F.J., (1991), "Public Safety Issues in Collapse of L'Ambiance Plaza," Journal of Performance of Constructed Facilities, ASCE, Vol. 5, No. 2, pp. 92-112.

Cuoco, D., D. Peraza and T.Z. Scarangello, (1992), "Investigation of the L'Ambiance Plaza Building Collapse," Journal of Performance of Constructed Facilities, ASCE, Vol. 6, No. 4, pp. 211-231.

McGuire, W., (1992), "Comments on L'Ambiance Plaza Lifting Collar/Shearheads," Journal of Performance of Constructed Facilities, ASCE, Vol. 6, No. 2, pp. 78-95.

Moncarz, P.D., R. Hooley, J.D. Osteraas and B.J. Lahnert, (1992), "Analysis of Stability of L'Ambiance Plaza Lift-Slab Towers," Journal of Performance of Constructed Facilities, ASCE, Vol. 6, No. 4, pp. 232-245.

L'Ambiance Plaza Collapse

BURNABY SUPERMARKET ROOFTOP PARKING DECK - (1988)

Part of the rooftop parking deck of a supermarket at the Station Square development in Burnaby, British Columbia, Canada, collapsed on opening day, during a Grand Opening ceremony on April 23, 1988. The ceremony was a special preview opening for neighborhood senior citizens. They had been directed to park on the 69 by 122 m (225 by 400 ft), one-story building, which was part of a regional community shopping center that included a hotel, apartments, retail space, theaters, offices and the flagship building for the Save-on-Foods retail chain. After a welcoming program, the 600 senior citizens began to shop, aided by 370 employees of the Save-on-Foods store.

About 15 minutes later a loud bang was heard, and water began to flow from a broken overhead fire sprinkler pipe. A photograph was taken of the broken pipe and a severely distorted beam-to-column connection. After this distortion and pipe rupture, the supermarket staff acted promptly and efficiently to clear people from the immediate area and then to begin evacuation of the entire store. Approximately 4-1/2 minutes later the roof in four bays collapsed into the shopping area, along with 20 automobiles. The collapsed area was 27 by 23 m (87 by 75 ft). No one was killed, but 21 people were injured.

A commissioner was appointed two weeks after the failure by the government of British Columbia to investigate the failure. The commissioner's report detailed the probable technical cause of the collapse and also reviewed the many procedural deficiencies that led to this failure. The recommendations included in the report continue to effect important revisions to standards of practice in Canada.

The 8,400 sq m (90,000 sq ft) supermarket building is a single-story steel column and beam structure that uses the cantilevered or Gerber beam system. Wide-flange steel beams pass over hollow steel tube columns, extending as cantilevers at each end. The beams support open-web steel joists, which, in turn, support a composite concrete and corrugated metal deck. The most probable technical cause of the failure was insufficient stability of the beam-to-column connection. There was no provision in the design for lateral support to the bottom flange of the beam, at a condition of bending moment that placed this flange in compression. This type of failure has occurred in the past, and the need to investigate this mode of failure has been well-established.

Lessons Learned:

Of greater interest than the technical cause of this failure are the many procedural deficiencies in the project delivery system that permitted the design deficiency to go unrecognized. The commissioner's report cites numerous contributing procedural problems, including competitive bidding for design services, unclear assignment of responsibilities, inadequate involvement of designers during the construction phase, poorly-monitored changes during construction, incomplete peer review, and inadequate professional liability insurance. These procedural problems are certainly not unique to Canada, but are the same set of deficiencies that plague the construction industry in the United States. The commissioner's report contained 19 recommendations, including 1) independent project peer review funded by increased permit fees; 2) special examinations for structural engineers and mandatory professional liability insurance; 3) the development by the provincial government of a manual that would clarify the responsibilities of all parties to the construction process; 4) a minimum fee schedule for design services; and 5) strengthening of certain steel industry guidelines and design manuals, particularly with respect to beam-column connection support requirements.

References:

Galambos, T. V., ed., (1987), "Guide to Stability Design Criteria for Metal Structures", 4th edition, John Wiley and Sons, New York, NY.

Report of the Commissioner Inquiry - Station Square Development, (1988), Government of British Columbia, Victoria, B. C., Canada.

Jones, C. P. and N. D. Nathan, (1990), "Supermarket Roof Collapse in Burnaby, B.C., Canada", Journal of Performance of Constructed Facilities, ASCE, Vol. 4, No. 3, pp. 142-160.

Baer, B. R., (1992), "Discussion: Supermarket Roof Collapse in Burnaby, B.C., Canada, Journal of Performance of Constructed Facilities, ASCE, Vol. 6, No. 1, pp. 67-68.

Jones, C. P. and N. D. Nathan, (1992), "Closure To Discussions: Supermarket Roof Collapse in Burnaby, B.C., Canada", Journal of Performance of Constructed Facilities, ASCE, Vol. 6, No. 1, pp. 69-70.

Burnaby Supermarket Rooftop Parking Deck Collapse

NORTHRIDGE MEADOWS APARTMENTS - (1994)

A large proportion of the most severe damage to buildings in the January 17, 1994, Southern California earthquake was experienced by condominium buildings. Several thousand of these structures were situated in the most severely shaken areas and the largest single incidence of fatalities, sixteen individuals, occurred in one such building, the Northridge Meadows Apartments.

The typical configuration of these buildings provides a lowest level parking area, either of timber framed carport configuration or comprising a reinforced concrete structural slab supported on perimeter walls and internal columns, above which is constructed two or three levels of stucco clad, timber framed, residential units.

A consistent pattern of observed seismic weaknesses was noted. These include overloading of the perimeter walls, incipient punching shear damage of column/slab connections, in-plane shear failures of stucco and drywall clad walls, cracking of vertical timber studs, splitting of bottom sill plates, shattering of lightweight concrete floor slabs, out-of-plane separation of stucco and many connection detail deficiencies.

A particularly interesting aspect of the behavior of the condominium buildings in the Northridge earthquake is that many clearly experienced a seismic load demand that approximated closely to the available capacity. Consequently the response observations provided an invaluable data base with which to assess the effectiveness of the design and construction practices followed.

In the case of the Northridge Meadows Apartments, the collapse of several blocks can be ascribed directly to the soft bottom story, as a consequence of insufficient shear resistance being provided at this level.

Lesson Learned:

The style of multi story, multi family residential buildings typified by the Northridge Meadows Apartments proved unacceptably hazardous in the moderate to strong earthquake shaking experienced on January 17, 1994. The frailties of the structural form were exposed. Stricter code provisions and more stringent inspection procedures to ensure improved quality control of the construction process will reduce the possibility of a repetition of this poor performance in the future.

References:

"Preliminary Report of the Seismological and Engineering Aspects of the January 17, 1992, Northridge Earthquake", (1994), Report UCB/EERC 94/01, Earthquake Engineering Research Center, University of California at Berkeley.

Hall, John. F. (Editor), (1994), "Northridge Earthquake, January 17, 1994; Preliminary Reconnaissance Report", Earthquake Engineering Research Institute, Oakland, California.

Northridge Meadows Apartments Collapse

CAL. STATE UNIV., NORTHRIDGE, OVIATT LIBRARY - (1994)

The Magnitude 6.4 January 17, 1994 Southern California earthquake, although not a major event in terms of total energy release, was relatively shallow with its epicenter in a developed urban area. Consequently the local ground shaking was very intense, the damage to constructed facilities was severe and the economic loss was huge. First indications were that only minor damage had been sustained by steel framed structures. Once such building was the four story steel braced frame Oviatt Library building on the campus of California State University, Northridge. In common with many flexible structures, architectural finishes were found to have been damaged but this was ascribed to the inherent incompatibility of the relatively flexible steel skeleton and the stiffer cladding. The observation that substantially greater, readily apparent, architectural damage occurred in some of the aftershocks resulted in a more thorough inspection being undertake. This revealed fractures in the steel column base plates and in various other welded connection in the steel framework.

Since steel framed structures comprised the seismic resisting system of choice for tall buildings, the established fact that this public building had not performed satisfactorily, coupled with unconfirmed reports of several unexpected failures of other steel frames, prompted a rigorous inspection of all the buildings falling into this category in the zone of strong to moderate ground shaking.

It was found that in more than one hundred buildings brittle fractures had occurred in welded connections, including the moment resisting joints between beams and columns. Detailed analyses indicated that these buildings had been designed and constructed in conformity with normal industry standards.

The disturbing conclusion drawn from these observations was that structural engineers could no longer have confidence in the procedures used to design and construct steel framed buildings in earthquake prone areas, notwithstanding that these buildings were in compliance with the appropriate codes. Additionally, of even greater concern in the short term, no consensus existed as to an acceptable procedure for repair of the damaged frames.

Faced with an exceptionally challenging compound technical problem, those concerned with the re-establishment of steel moment resistant frame buildings as a viable alternative amongst the several structural choices available to earthquake engineers, cooperated to devise and initiate a program with the focused objective of leading to the development of standards for the repair, retrofit and design of steel moment resisting frame buildings so that they will provide reliable, cost-effective performance in future earthquakes.

The experience of the directed program of professional practice development coupled with problem focused investigations is likely to be reviewed with great interest by those who may wish to consider the applicability of this approach to other engineering challenges.

Lessons Learned:

The Northridge earthquake destroyed the faith of the structural engineering profession in the seismic resisting system of choice for tall buildings, namely a steel frame with moment resisting beam-column.

With hindsight it appears that the problems in steel buildings exposed by the Northridge earthquake could have been anticipated by a prudent analyzer of the configuration of many modern steel buildings and of the moment resisting joints which have become industry standard over the last ten years.

However, as is so often the case with technological progress, the pressures to adopt new, supposedly more cost efficient, techniques outran the performance verification process necessitating major reconsideration of the recently accepted industry standards.

References:

Hall, John. F. (Editor), (1994), "Northridge Earthquake, January 17, 1994; Preliminary Reconnaissance Report", Earthquake Engineering Research Institute, Oakland, California.

AISC, (1994), "Localized Steel Damage", American Institute of Steel Construction Inc., Modern Steel Construction, 34 (4).

CSSC, (1994), "Damage to Steel Frame Buildings by the Northridge Earthquake", Public Advisory issued by the California Seismic Safety Commission, Sacramento, California.

Bertero, V.V., Anderson, James, C. & Krawinkler, H., (1994), "Performance of Steel Building Structures during the Northridge Earthquake", Report No. UCB/EERC-94/09, Earthquake Engineering Center, University of California, Berkeley.

SAC, (1994), "Program to Reduce Earthquake Hazards in Steel Moment Frame Structures", formulated by the SAC Joint Venture Partnership, Structural Engineers Association of California, Sacramento, California.

SAC, (1994), "Steel Moment Frame Connection, Advisory No. 1", SAC Joint Venture Partners, Structural Engineers Association of California, Sacramento, California.

City of Los Angeles, (1994), "Repairs of Cracked Moment Frame Connections in Welded Steel Frame Structures and Requirements for SMRF-Welded Connections in New Buildings", Plan Check Information.

AISC, (1994), "Northridge Steel Update 1", American Institute of Steel Construction Inc., Chicago, Illinois.

Oviatt Library Roof Failure

INDEX